Lecture Notes in Physics

Edited by J. Ehlers, München, K. Hepp, Zürich
R. Kippenhahn, München, H. A. Weidenmüller, Heidelberg
and J. Zittartz, Köln
Managing Editor: W. Beiglböck, Heidelberg

147

J. Messer

Temperature Dependent Thomas-Fermi Theory

Springer-Verlag
Berlin Heidelberg New York 1981

Author

Dr. Joachim Messer
Universität München, Sektion Physik
Theoretische Physik
Theresienstraße 37, D-8000 München 2

ISBN 3-540-10875-0 Springer-Verlag Berlin Heidelberg New York
ISBN 0-387-10875-0 Springer-Verlag New York Heidelberg Berlin

CIP-Kurztitelaufnahme der Deutschen Bibliothek

Messer, Joachim:
Temperature dependent Thomas-Fermi theory / J. Messer. – Berlin; Heidelberg;
New York: Springer, 1981.
(Lecture notes in physics ; 147)
ISBN 3-540-10875-0 (Berlin, Heidelberg, New York);
ISBN 0-387-10875-0 (New York, Heidelberg, Berlin)
NE: GT

Printing and binding: Beltz Offsetdruck, Hemsbach/Bergstr.
2153/3140-543210

PROLOGUE

These lecture notes had their origin in a course given in the framework of the "Interuniversitaire Derde Cyclus Programma 'Veldentheorie en Statistische Mechanika'" during the winter term of the academic year 1979/80 at the Instituut voor Theoretische Fysica of the Universiteit Leuven and from lectures given at the Universities of Göttingen, Groningen, Munich and Vienna.

The choice of the subjects was made according to physical relevance and mathematical rigour. In this respect we followed the guidelines formulated, e.g. in [1]. Consequently a certain amount of literature, prescriptions and computations, which has been performed in the field of hot Thomas-Fermi models, had to be dropped from presentation.

The author is grateful for discussions, suggestions, remarks during the lectures, and correspondence to Profs. and Drs. B. Baumgartner, H.J. Borchers, R. Dekeyser, M. Fannes, W.D. Garber, J. Lopuszanski, D. Lynden-Bell, H. Narnhofer, A. Pflug, H. Satz, G.L. Sewell, H. Spohn, W. Thirring, A. Verbeure, and J. Yngvason.

In particular I am greatly indebted to Professor H.J. Borchers for teaching me mathematical physics and for everything I was able to learn in many years.

I am also indebted to Professor W. Thirring for directing my interest to Thomas-Fermi theory of cosmic matter during my stay at the University of Vienna in summer 1978.

It is a great pleasure to thank Professor A. Verbeure for the generous hospitality extended to me at Leuven, for many illuminating discussions and for his interest in Thomas-Fermi problems.

Munich, June 1981 J. MESSER

"Die Thomas-Fermi Theorie von Atomen und Sternen stellt das einzige Vielkörperproblem mit realistischen Kräften dar, welches im entsprechenden thermodynamischen Limes der menschlichen Rechenkunst erlegen ist."

W. Thirring [2]

CONTENTS

INTRODUCTION

The structure of matter is based on the properties of electrons and atomic nuclei. Their interactions are mostly electrostatic and gravostatic, due to their nonzero charges and masses. If the gravostatic forces can be regarded as small compared to the electrostatic forces, we speak of "ordinary matter", as we are faced with this limiting case in our (ordinary) terrestrial laboratories, whereas in the opposite case we use the notion of "purely cosmic matter" referring to its occurrence in stellar objects in the universe, the cosmic laboratory. Mainly we shall discuss both interactions on the same footing, and use the term "ordinary cosmic matter" for this general description of matter. Ordinary cosmic matter consists of one sort of charged fermions with masses M_1 ("electrons"), interacting by gravitational and electric forces with another sort of oppositely charged fermions with masses M_2 ("nuclei"). The total amount of charge carried by a nucleus is Z times the charge carried by an electron, with $Z \geq 1$, $Z \in \mathbb{N}$. A matter model consisting of electrons, protons and neutrons only ($Z = 1$) seems not to be satisfactory, because for helium or heavier elements the protons are tightly bound in the nucleus by strong nuclear forces, despite of their Coulomb repulsion, and these strong interactions (as well as the weak interactions) are not contained in the Hamiltonian. A system of electrons, neutrons, and protons with electrostatic and gravostatic interactions cannot lead to the formation of nuclei (through nuclear forces) and normal matter. Therefore we introduce the nuclei from the beginning. Although the nuclei may have different statistics, we confine ourselves to fermions, because the Thomas-Fermi theory for bosons is not yet well understood (see Epilogue). A further simplification will be the neglection of relativistic effects, which we shall regard as small corrections, as long as the velocity of the electrons is small compared to the speed of light. This is even true in the interior of supernovae.

The particle numbers in ordinary matter are typically of the order of 10^{24} and gravitation dominates electrostatics for particle numbers above 10^{54}. To discuss the bulk thermodynamic properties of such large systems, we shall consider infinite systems, the limit systems of a sequence of finite systems with increasing particle numbers. We expect most thermodynamic properties to be close to those of the infinite system. Considering infinite matter also helps us in defining phase transitions as non-analytic behaviour of thermodynamic functions, whereas how to define a phase transition for a finite system is not clear, because in this case the thermodynamic functions are analytic. If the interaction energy is extensive, more precisely, bounded from below proportional to the particle number, then there is a well-known usual way for performing the thermodynamic limit. In presence of gravitation the interaction energy turns out to be non-extensive and the usual procedure for the thermodynamic limit becomes incorrect. The well-founded thermodynamic limit has to take into account the peculiar scaling behaviour for "non-stable" interactions, like gravitation. We call this limit the "thermodynamic Thomas-Fermi limit". It is the adequate thermodynamic limit for non-extensive systems. In this limit the Thomas-Fermi hypothesis, and the single particle picture become exact. There are two exactly solvable cases: The one electron hydrogen atom, and the Thomas-Fermi atom, which describes the exact quantum mechanics for infinitely many electrons.

Statistical quantum systems are completely described by stating all probabilities for finding the particles at prescribed places. The mathematical notions for these probabilities are the correlation functions and more general the "states" on a given algebra of observables. In the thermodynamic Thomas-Fermi limit we turn from a microscopic system, governed by the laws of quantum mechanics, to a macroscopic system subject to the laws of classical physics. Consequently there appears a macroscopic scale (sometimes called hydrodynamic) besides the microscopic scale and a macroscopically quasi-local algebra (the "hydrolocal algebra") besides the usual microscopically quasi-local algebra. The latter object has proved its utility since a long time, the former was introduced recently and is very useful because states on the hydrolocal algebra might show quantum correlations between macroscopically separated positions.

In case there is coexistence of phases the thermodynamic limit of the equilibrium state is a convex combination of extremal states, which describe the pure phases. In this case there exist at least two solutions of the Thomas-Fermi equation, because the correlation functions are uniquely given by the solutions of the Thomas-Fermi equation in the limit. This is also visible in the non-analytic behaviour of the limiting free energy. If there are two or more solutions of the Thomas-Fermi equation, then the free energy is not everywhere differentiable with respect to the temperature, showing an Ehrenfest phase transition of the first kind. Such a non-analytic behaviour is shown to be typical for (purely) cosmic matter where the phase transition, induced by gravitational attraction, turns an almost Boltzmann gas into a degenerate Fermi gas as condensate and vice versa. After the implosion the condensate remains stable, because of the Fermi pressure of the nonrelativistic particles. In the relativistic case the Fermi pressure could not stabilize the implosion if a certain particle number (the "Chandrasekhar limit") is exceeded, and a complete collapse would result. The phase transition for cosmic matter is related to the supernova phenomenon and the birth of the neutron star, which corresponds to the condensate. There is no phase transition in the microcanonical ensemble, but a region of negative specific heat, and gravitating fermions therefore have inequivalent ensembles.

I. *THE THERMODYNAMIC THOMAS-FERMI LIMIT FOR THERMODYNAMIC FUNCTIONS*

I.a. Ground-State Energy

I.aa. Stability of Ordinary Matter (Coulomb Systems)

As a pre-study to the thermodynamics of matter we investigate first the dependence of the energy on the particle number N. This gives us the scaling behaviour, which enters the definition of the thermodynamic limit. In subsection aa. we discuss bounds to the ground-state energy for ordinary matter of N_t particles, with N electrons and K nuclei, and in subsection ab. we treat the same problem for cosmic matter of N neutral particles.

Ordinary matter is to be understood as a system of charged electrons which are fermions and oppositely Z-times charged nuclei with arbitrary statistics, interacting via Coulomb forces.

The extensive nature of the energy of real matter at laboratory sizes is considered as a fundamental property. It is an empirical fact that the binding energy per particle in bulk matter is not essentially bigger than in isolated atoms. The chemical forces saturate and bulk matter does not condensate in a way such that electrons are close to nuclei. This stability property is deducible from the many-body Hamiltonian H_N for ordinary matter with N electrons, and K nuclei, and the total particle number $N_t = N + K$. We formulate the stability problem in the following mathematically precise way: Find an $A > 0$, independent of the total particle number N_t, such that the ground-state energy $E_N = \inf (\psi, H_N \psi) \geq -A\, N_t$, where the inf is taken over all wave functions in the domain of H_N which are antisymmetric in the electron coordinates. We emphasize that N_t is the total number of particles, not to be confused with the number N of electrons or the number of nuclei K. The extensivity of the energy means that considering only half of all particles by dividing the system gives two subsystems each of which carry half the energy. We express stability by a lower bound to the energy which is extensive, thus preventing the subsystems to lump together, which would be energetically more favourable otherwise. The energy A is (best possible) of the order of the binding energy per particle in atoms, e.g. for hydrogen $A \simeq 1$ Ry $\simeq 13.6$ eV.

To obtain the lower bound $-A\, N_t$ to the many-body Hamiltonian H_N, it is essential that one sort of particles (e.g. the electrons) are fermions. The statistics of the other sort (e.g. the nuclei) does not matter. Deuterium or helium is as stable as hydrogen. In the Coulomb Hamiltonian a lot of cancellations occur in order that the sum of (approximately) N^2 unbounded terms grow only proportional to N. These cancellations are due to both, the alternating sign of the potentials, describing attraction and repulsion according to the charges alternatingly, as well as the Fermi statistics for the electrons. For example, if both sorts of particles would be bosons, then there exist A,B > O (independent of N_t), such that the cancellations are less miraculous [3,4]:

$$(1) \qquad -B\, N_t^{5/3} \leq E_N \leq -A\, N_t^{7/5} \ .$$

Therefore bosons are not stable and the Pauli principle is crucial for stability. For bosons a common upper and lower bound to the groundstate energy has not yet been proven. It is conjectured to be $N^{7/5}$ [3, 4,5], however in case of infinitely heavy nuclei Lieb proved the validity of the $N^{5/3}$ upper bound [6]. If the $N^{7/5}$ conjecture would be correct, this result means that the limits $N \to \infty$ and the mass of the nuclei $\to \infty$ are not interchangeable.

The stability of the Coulomb Hamiltonian is - even for bosons - easily achieved by cutting off the singularity at the origin [7]. If at short distances the Coulomb potential is replaced, e.g. $1/r \to (1-\exp(-\mu r))/r = v_\mu(r)$, then even the interaction energy becomes stable. To see this, notice that $v_\mu(r)$ is a function of positive type, i.e. for all (complex numbers) z_i

$$(2) \qquad \sum_i \sum_j z_i z_j^* v_\mu(x_i - x_j) \geq 0.$$

Therefore the interaction energy has a lower bound $-N_t \frac{1}{2} \mu$. However the cuttoff μ would be due to the size of the nuclei and the obtained binding energy per particle would be in the range of MeV and does not describe stability of matter properly. It is therefore important not only to solve the mathematical problem of deducing a lower bound proportional to N_t from the first principle Hamiltonian H_N for any constant A, but also to obtain a best possible constant A which is in the range of eV.

The stability problem for ordinary matter was first solved by Dyson and Lenard [3,4]. They obtained a value of approximately 10^{14} Ry for A.

A simplified proof was then given by Lieb and Thirring [8] with $A \simeq 33.28$ Ry (computed for neutral hydrogen). This value was further improved by Lieb [9]. Recently Thirring [10] obtained a considerable improvement for large nuclear charges Z which is "best possible" in the limit $Z \to \infty$ (then $A \simeq 1.54\, Z^{7/3}$ Ry).

For a detailed introduction to the stability theory for matter we refer to the excellent reviews by Lieb [11] and by Thirring [12]. In the following we shall briefly sketch some aspects of the Lieb-Thirring stability considerations.

I.aa.1. Sobolev and Lieb-Thirring Inequalities

Consider one particle in three dimensions with wave function $\psi \in \mathcal{L}_2(\mathbb{R}^3)$ and with kinetic energy T_ψ and probability density $\rho_\psi(x)$:

$$T_\psi = \int |\nabla\psi(x)|^2 d^3x, \tag{3}$$

$$\rho_\psi(x) = |\psi(x)|^2 \quad , \quad \|\rho_\psi\|_1 = 1 . \tag{4}$$

The "Sobolev Inequality" states

$$T_\psi \geq K_s \|\rho_\psi\|_3 , \tag{5}$$

with the best possible constant $K_s = 3(\frac{\pi}{2})^{4/3}$. For a plausible derivation see [11,13], for the proof see [14]. The Sobolev inequality relates the kinetic energy and an expression involving the localization probability. Consequently it represents a version of the common uncertainty principle. This alone suffices to demonstrate the stability of single atoms, i.e. finding a lower bound to the ground-state energy of an atom, consisting of N negatively charged electrons, carrying the unit charge e and one nucleus of positive charge $Z|e|$. This lower bound is proportional to the particle number $N + 1$, showing extensivity of the energy. Placing the nucleus at the origin and choosing units in which $\hbar = 2m = |e| = 1$, with m being the electron mass, we write the formal Hamiltonian of the atom as

$$H_N = \sum_{i=1}^{N} (-\Delta_i - Z|x_i|^{-1}) + \sum_{1\le i<j\le N} |x_i - x_j|^{-1} \geq \sum_{i=1}^{N} h_i , \tag{6}$$

with

$$h_i = -\Delta_i - Z|x_i|^{-1} . \tag{7}$$

It is now easy to see that

$$H_N \geq N\, E_o(Z) , \tag{8}$$

if the ground-state energy of the hydrogen atom has a finite lower bound $E_o(Z)$. Applying the Sobolev inequality (5) to the hydrogen Hamiltonian (7) (with the index i dropped) leads to the functional $h[\rho]$:

$$(\psi,h\psi) \geq K_s\Big(\int \rho_\psi^3(x)\,d^3x\Big)^{1/3} - Z\int |x|^{-1}\rho_\psi(x)\,d^3x =: h[\rho] \tag{9}$$

with $\|\psi\|_2 = 1$. The minimum of $h[\rho]$ with respect to ρ under the subsidiary conditions $\rho(x) \geq 0$, $\|\rho\|_1 = 1$ is attained at some ρ_o [see e.g. 11] and

$$h[\rho_o] = -\frac{4}{3} Z^2\, Ry = E_o(Z) , \tag{10}$$

which is close to the correct value for the ground-state energy. This simple calculation shows the utility of the Sobolev inequality for stability considerations. We want to stress the fact that the Sobolev inequality alone (as uncertainty relation) does not suffice to prove stability in the case where an indefinite number of nuclei is present; then Fermi statistics is needed for one sort of the particles. For a single atom, however, the above simple calculation holds for every statistics. Stability of single atoms holds also for bosons. The worst configuration which can occur is placing all bosonic electrons in the ground-state of the Coulomb field of the single nucleus, which leads an extensive energy. If the electrons of a single atom are fermions a state, lower than the ground state, is obtained by filling the energy levels of the hydrogen spectrum according to the degeneracies. This leads $\frac{1}{4}\, 3^{4/3}(N+1)^{1/3}E_o(Z)$ as lower bound to the Hamiltonian. Atoms show better stability properties than ordinary matter.

By Hölder inequality

$$(11) \qquad \int \rho(x)^{5/3} d^3x \leq \left(\int \rho(x)^3 d^3x\right)^{1/3} \left(\int \rho(x) d^3x\right)^{2/3} ,$$

one derives a weaker version of the Sobolev inequality:

$$(12) \qquad T_\psi \geq k \int \rho_\psi(x)^{5/3} d^3x ,$$

with $k = K_s$, which was the best possible constant for (5), but is not best possible for (12) [see e.g. 11]. (12) still holds if $k = K^c = (\frac{3}{5})(6\pi^2)^{2/3}$. Using $k = K^c$ instead of K_s and (12), the above minimization for atoms can be repeated, leading the new lower bound $E_o'(Z)$ instead of $E_o(Z)$ with

$$(13) \qquad E_o'(Z) = -3^{1/3}\, Z^2\, Ry \simeq 1.082\, E_o(Z).$$

This demonstrates that also the weakened version (12) of the Sobolev inequality might be extremely useful for stability calculations. Therefore we extend (12) to the N body fermion case, arriving at the Lieb-Thirring inequality.

Let $x_i \in \mathbb{R}^3$ denote the positions for the ith electron $i \in \{1,2,\ldots,N\}$ with spin variable $\sigma_i \in \{1,2,\ldots,q\}$ and let the wave function be $\psi(x_1,\ldots, x_N; \sigma_1,\ldots,\sigma_N)$, being antisymmetric in the pairs (x_i,σ_i) with finite norm

$$(14) \qquad (\psi,\psi) = \sum_{\sigma_1=1}^{q} \cdots \sum_{\sigma_N=1}^{q} \int |\psi(x_1,\ldots,x_N;\sigma_1,\ldots,\sigma_N)|^2 d^3x_1 \ldots d^3x_N$$

on $\mathcal{L}_2(\mathbb{R}^{3N},\mathbb{C}^{9N})$. Let

$$(15) \qquad T_\psi = N \sum_{\sigma_1=1}^{q} \cdots \sum_{\sigma_N=1}^{q} \int |\nabla_1 \psi(x_1,\ldots,x_N;\sigma_1,\ldots,\sigma_N)|^2 d^3x_1 \ldots d^3x_N,$$

$$(16) \qquad \rho_\psi(x) = N \sum_{\sigma_1=1}^{q} \cdots \sum_{\sigma_N=1}^{q} \int |\psi(x,x_2,\ldots,x_N;\sigma_1,\ldots,\sigma_N)|^2 d^3x_2 \ldots d^3x_N,$$

then

$$(17) \qquad T_\psi \geq (4\pi q)^{-2/3} K^c \int \rho_\psi(x)^{5/3} d^3x.$$

The inequality (17), which was derived in [8], is the N fermion pendant to the weakened Sobolev inequality (12) and is called "Lieb-Thirring Inequality". It is conjectured [11] that the constant $(4\pi q)^{-2/3} K^c$ can be improved to $q^{-2/3} K^c$. Lieb achieved an improvement by a factor 1.496 [9].

The proof of (17) is based on

Lemma (18): Suppose $V(x) \leq 0$ and $V \in \mathcal{L}_{5/2}(\mathbb{R}^3)$. If $E_1 \leq E_2 \leq \ldots \leq E_\ell \leq 0$ are the negative eigenvalues of the Schrödinger operator $H = -\Delta + V(x)$, then

$$(19) \qquad \sum_{j=1}^{\ell} |E_j| \leq \frac{4}{15\pi} \int |V(x)|^{5/2} d^3x \,.$$

For a proof of Lemma (18) see [8].

Proof of the Lieb-Thirring inequality (17):

Choose in Lemma (18) $V(x) = -\alpha\, \rho_\psi(x)^{2/3}$ with $(\frac{2}{3\pi}) q\alpha^{3/2} = 1$ for a fixed $\psi \in \mathcal{L}_2(\mathbb{R}^{3N},\mathbb{C}^{qN})$ being antisymmetric in the pairs (x_i,σ_i). Consider the Hamiltonian $\tilde{H}_N = \sum_{i=1}^{N} -\Delta_i + V(x_i)$ with the fermion ground-state energy E_o. If $N \leq \ell q$, where ℓ is, as in Lemma (18), the number of all negative eigenvalues, then $E_o \geq q \sum_{j=1}^{\ell} E_j$, by filling the lowest energy levels with q electrons each. If $N > \ell q$, the lowest energy possible for fermions is obtained if the levels are filled as above and the $N - \ell q$ surplus particles have wave functions located far away from the origin carrying an arbitrary small amount of kinetic energy. Then their contribution to the energy is arbitrarily small and $E_o \geq q \sum_{j=1}^{\ell} E_j$ holds. Since

$$(20) \qquad E_o \leq (\psi, \tilde{H}_N \psi) = T_\psi - \alpha \int \rho_\psi(x)^{5/3} d^3x$$

the proof is concluded using (19). □

I.aa.2. Ground-State Thomas-Fermi Theory as Mathematical Method

The full Hamiltonian for ordinary matter is trivially bounded from below by the Hamiltonian H_N for N electrons, K static nuclei at locations $R_1, \ldots, R_K$ with charges $z_1, \ldots, z_K$. Again we choose $\hbar = 2m = |e| = 1$ (with m being the electron mass) as units. This Hamiltonian H_N for the fixed nuclei writes formally:

$$(21) \qquad H_N = \sum_{i=1}^{N} - \Delta_i - V(x_i) + \sum_{1 \leq i < j \leq N} |x_i - x_j|^{-1} + U(\{z_j, R_j\}),$$

$$(22) \qquad V(x) = \sum_{j=1}^{K} z_j |x - R_j|^{-1} ,$$

$$(23) \qquad U(\{z_j, R_j\}) = \sum_{1 \leq i < j \leq K} z_i z_j |R_i - R_j|^{-1} .$$

Stability of ordinary matter is proved, if we can find a lower bound to H_N in (21), proportional to $N_t = N + K$ with a constant independent of the locations of the nuclei. The second main ingredient in this proof is the Thomas-Fermi theory for the ground-state, i.e. at temperatures $T = 0$. It will be used as a mathematical method in obtaining lower bounds to Coulomb-like Hamiltonians. In this subsection we give a brief description of some aspects of ground-state Thomas-Fermi theory. For more details the interested reader is referred to the references.

The Thomas-Fermi energy functional on the convex, proper cone

$$\mathcal{S} = \{\rho\in\mathcal{L}_{5/3}(\mathbb{R}^3)\cap\mathcal{L}_1(\mathbb{R}^3)\ ,\ \rho \geq 0\}$$

is given by:

$$E[\rho] = q^{-2/3}\ K^c\int\rho(x)^{5/3}d^3x - \int V(x)\rho(x)d^3x + \tag{24}$$

$$+ \frac{1}{2}\int\rho(x)|x-y|^{-1}\rho(y)d^3xd^3y + U(\{z_j,R_j\})\ .$$

It has the following properties [15]: the map $\mathcal{S}\to\mathbb{R}$ ($\rho\to E[\rho]$) is

(25) bounded from below,

(26) on each in $\mathcal{L}_{5/3}\cap\mathcal{L}_1$ bounded sets lower semicontinuous in the weak topology of $\mathcal{L}_{5/3}$,

(27) strictly convex,

(28) Gâteux differentiable.

The Thomas-Fermi energy of $\lambda\in\mathbb{R}_+$ electrons exists:

$$E_\lambda^{TF} = \inf\ \{E[\rho]/\rho\in\mathcal{S}\text{ and }\|\rho\|_1 = \lambda\}\ . \tag{29}$$

The variational or Euler-Lagrange equation to (24), (29) is

$$\frac{5}{3}\ K^c\ q^{-2/3}\ \rho^{2/3}(x) = \max\{\phi[\rho](x) - \mu, 0\}\ \text{with} \tag{30}$$

$$\phi[\rho](x) = V(x) - \int\rho(y)|x-y|^{-1}d^3y \tag{31}$$

and is called the Thomas-Fermi equation. It has a solution with $\|\rho\|_1 = \lambda$ if and only if there is a minimizing ρ for E_λ^{TF}. For the ground-state

Thomas-Fermi theory the following properties are known:

<u>Theorem (32)</u> [16,17]: Let $\lambda \leq Z = \sum_{j=1}^{K} z_j$, then

(33) $E[\rho]$ attains its minimum at an element ρ_λ^{TF} of the set

$$\mathcal{S}_\lambda = \{\rho \in \mathcal{S} / \|\rho\|_1 = \lambda\}.$$

(34) The ρ_λ^{TF} is unique and a solution of the Thomas-Fermi equation.

(35) The chemical potential is

$$-\mu = \frac{\partial E_\lambda^{TF}}{\partial \lambda} \leq 0$$

(36) $\mu = 0$ if and only if $\lambda = Z$.

(37) $\lambda \to \mu$ is a continuous, convex, decreasing, and surjective function from $(0,Z]$ onto $[0,\infty)$.

(38) For all $x \in \mathbb{R}^3$, $\lambda \in [0,Z]$: $\phi[\rho_\lambda^{TF}](x) > 0$.

Let now $\lambda > Z$ (negative ions), then

(39) $E[\rho]$ does not attain its minimum on $\mathcal{S}_\lambda$. The Thomas-Fermi equation has no solution on $\mathcal{S}_\lambda$.

<u>Proof</u>: See [16,17].

The ground-state Thomas-Fermi theory is the exact quantum mechanics of the infinite electron atom, i.e. it becomes exact in the appropriate limit where the nuclear charge $Z \to \infty$ [16,18,17]. The Thomas-Fermi energy E_λ^{TF} is always larger or equal to E_Z^{TF} with $Z = \sum_{j=1}^{K} z_j$ and turns out to be [11]

$$E_\lambda^{TF} \geq E_Z^{TF} \geq -(2.21) q^{2/3} (K^c)^{-1} \sum_{j=1}^{K} z_j^{7/3} \quad . \tag{40}$$

Finally we need the following

<u>Lemma (41)</u>: Let $(x_1,\ldots,x_N) \in \mathbb{R}^{3N}$ have non-coincident components and let

(42) $$V_X(y) = \sum_{j=1}^{N} |y-x_j|^{-1} \ , \ \gamma > 0, \text{ and } \rho \in \mathcal{S} \ .$$

Then

(43) $$\sum_{1\leq i<j\leq N} |x_i-x_j|^{-1} \geq -\frac{1}{2}\int \rho(x)\,|x-y|^{-1}\rho(y)\,d^3x\,d^3y - (2.21)N\,\gamma^{-1} + \int \rho(y)V_X(y)\,d^3y - \gamma\int \rho(y)^{5/3}d^3y \ .$$

<u>Proof</u>: (43) is the inequality $E[\rho] \geq E_\lambda^{TF}$ together with (40) if one fixes in (24) $q = 1$, $K = N$, $z_j = 1$, $R_j = x_j$ for all $j = 1,\ldots,N$ and replaces K^c by γ. This involves that K^c is substituted by γ in (40). □

Teller [19] proved that there is no chemical binding in the framework of Thomas-Fermi theory. His proof was completed to mathematical rigour in [16,17]. The "no-binding theorem" of Teller asserts that the Thomas-Fermi energy of a molecule is always larger or equal to the sum of the energies of two pieces constituting the molecule:

(44) $$E_\lambda^{TF} \geq E_{\lambda_1}^{TF,1} + E_{\lambda-\lambda_1}^{TF,2} \ .$$

The Thomas-Fermi energy is unstable with respect to decomposition of molecules and to decomposition of a molecule into atoms. The estimates (40,43) are consequences of the no-binding theorem. We emphasize that the nucleon-nucleon Coulomb repulsion is essential. Neglecting U in (21) would reverse the inequality (44).

Stability of ordinary matter is now proved by attempting to show that up to a constant summand the Thomas-Fermi energy is a lower bound to the Hamiltonian H_N. Hereby the ground-state Thomas-Fermi theory appears to become a powerful mathematical method.

I.aa.3. <u>Stability of Bulk Matter</u>

In Sections I.aa.1 and 2. we have set up the necessary tools to prove the following theorem, which states the existence of an extensive lower bound to the N-particle Hamiltonian (21), being independent of the positions of the nuclei, thus delivering the desired stability bound to the Hamiltonian for ordinary matter.

Theorem (45) [8,11] : Let ψ be antisymmetric in the pairs (x_i, σ_i) and normalized, then for any $\gamma > 0$, such that α in

(46) $$\alpha = (4\pi q)^{-2/3} K^c - \gamma$$

is positive, we have

(47) $$(\psi, H_N \psi) \geq -(2.21) \left\{ N\gamma^{-1} + \alpha^{-1} \sum_{j=1}^{K} z_j^{7/3} \right\}, \text{ and}$$

(48) $$(\psi, H_N \psi) \geq -(2.21) \frac{(4\pi q)^{2/3} N}{K^c} \left\{ 1 + \left[\sum_{j=1}^{K} \frac{z_j^{7/3}}{N} \right]^{1/2} \right\}^2$$

$$=: \tilde{\tilde{E}}_N$$

$$\geq -(4.42)(4\pi q)^{2/3} (K^c)^{-1} \left\{ N + \sum_{j=1}^{K} z_j^{7/3} \right\}.$$

Proof: The electron-electron repulsion is estimated with the help of Lemma (41) to:

(49) $$\left(\psi, \sum_{1 \leq i < j \leq N} |x_i - x_j|^{-1} \psi \right) \geq \frac{1}{2} \int \rho_\psi(x) |x-y|^{-1} \rho_\psi(y) d^3x d^3y -$$

$$- (2.21)\, N\gamma^{-1} - \gamma \int \rho_\psi(x)^{5/3} d^3x$$

for any $\gamma > 0$, using

(50) $$(\psi, \{ \int \rho_\psi(y) V_x(y) d^3y \} \psi) = \int \rho_\psi(x) |x-y|^{-1} \rho_\psi(y) d^3x d^3y.$$

Applying the Lieb-Thirring inequality (17) for the kinetic energy, we arrive at

$$(51) \qquad (\psi, H_N\psi) \geq \alpha\int\rho_\psi(x)^{5/3}d^3x - \int V(x)\rho_\psi(x)d^3x + $$
$$+ \frac{1}{2}\int\rho_\psi(x)|x-y|^{-1}\rho_\psi(y)d^3xd^3y + U(\{z_j, R_j\}) - $$
$$- (2.21)\ N\gamma^{-1}\ .$$

Arbitrary γ's are permitted, as far as

$$(52) \qquad \alpha = (4\pi q)^{-2/3}\ K^c - \gamma$$

remains positive. Now Thomas-Fermi theory (with the constant $q^{-2/3}\ K^c$ replaced by α), in particular (40), and the fact that

$$(53) \qquad E_\alpha[\rho_\psi] \geq E^{TF}_{\alpha,N} = \inf\{E_\alpha[\rho]/\|\rho\|_1 = N\},$$

where $E_\alpha[\rho_\psi]$ is the Thomas-Fermi energy functional (24) with the constant α instead of $q^{-2/3}\ K^c$ applied to ρ_ψ, leads to the lower bound

$$(54) \qquad E_\alpha[\rho_\psi] \geq -(2.21)\ \alpha^{-1}\sum_{j=1}^{K} z_j^{7/3}\ .$$

The last inequality follows by using $[1 + \sqrt{a}]^2 \leq 2 + 2a$ (for any $a > 0$). If the nuclear charges z_i are bounded from above, then $\tilde{E}_N$ is bounded from below by the total particle number $N + K$. □

There is no assumption of neutrality made in (45) and matter may have a net charge. The fermions which we call "electrons" have q internal spin degrees of freedom. They coincide with the physical electrons if $q = 2$, they describe bosons if $q = N$. Then (48) proves the $N^{5/3}$ lower bound for bosons, already mentioned.

The scaling behaviour of the volume attained by ordinary matter is proportionality to N. The volume of ordinary matter is extensive. The mathematically precise formulation of volume extensivity is stated (without proof) in the following

Corollary (55) [8,11]: "Matter is bulky", i.e. for each $p \geq 0$, there exists $C_p > 0$

$$(56) \qquad (\psi, \sum_{i=1}^{N} |x_i|^p \psi) = \int |x|^p \rho_\psi(x) d^3x \geq C_p \, N \left(\frac{N^{5/3}}{|\tilde{E}_N|}\right)^{p/2} .$$

In particular, if $\sum_{j=1}^{K} \frac{1}{N} z_j^{7/3}$ is bounded, there exists $c_1 > 0$, such that the radius of the system behaves like

$$(57) \qquad (\psi, \frac{1}{N} \sum_{i=1}^{N} |x_i| \psi) > c_1 \, N^{1/3} ,$$

and c_1 is of the order a (a is the Bohr radius).

Proof: See [8,11].

Although proving the stability of (ordinary) matter means solving an eigenvalue problem of a linear partial differential equation, the Thomas-Fermi equation, formulated by physical intuition, turns out to be superior to the usual mathematical techniques. For further literature on stability of matter see e.g. [20,21,22].

I.ab. Saturation Properties of Cosmic Matter (Newton Systems)

Consider N neutral gravitating particles with Hamiltonian formally given by

$$(58) \qquad H_N = \sum_{i=1}^{N} - \frac{1}{2M_i} \Delta_i - \sum_{1 \leq i < j \leq N} \kappa M_i M_j |x_i - x_j|^{-1} .$$

κ denotes the (Newtonian) gravitational constant. Purely gravitational systems are expected to be less stable than ordinary matter, because they are one component systems with overall attraction. We study the scaling behaviour for energy and volume with respect to particle number first heuristically in several cases, and afterwards rigorously for gravitating fermions.

I.ab.1. Heuristic Considerations

1.1) Nonrelativistic Fermionic Newton Systems

The Hamiltonian is given formally by (58). Let all $M_i = M$ for simplicity. The Pauli principle for fermions can be used in the form of the "private room argument", introduced by Weisskopf. Each fermion occupies its "private room" with volume $(\Delta x)^3 = R^3/N$, where R is the linear dimension of the system. This distribution assures the fermions to keep distance of each other as much as possible. Then the Hamiltonian writes in crude approximation as

$$(59) \qquad H_N \simeq N \frac{p^2}{2M} - \frac{\kappa M^2}{R} \cdot \frac{N^2}{2} ,$$

$$(60) \qquad H_N \simeq N \cdot \frac{p^2}{2M} - \frac{\kappa M^2}{N^{1/3} \Delta x} \cdot \frac{N^2}{2} .$$

$$(61) \qquad H_N \simeq N \cdot \frac{p^2}{2M} - \frac{\kappa M^2}{2\hbar} p \cdot N^{5/3}$$

by the Heisenberg uncertainty relation $\Delta x \simeq \frac{\hbar}{p}$.
Minimize in (61) with respect to p, i.e.

$$(62) \qquad \frac{dH_N}{dp}(p_o) = 0 \rightarrow p_o = N^{2/3} \frac{M^3 \kappa}{2\hbar} .$$

Consequently the ground-state energy and the radius are

$$(63) \qquad H_N(p_o) \simeq E_N \simeq -N^{7/3} \frac{1}{4} \left[\frac{1}{2\hbar^2} M^5 \kappa^2 \right] =: -N^{7/3} \frac{1}{4} E_\kappa ,$$

$$(64) \qquad R \simeq R(\Delta x(p_o)) \simeq N^{-1/3} 2 \left[\frac{\hbar^2}{M^3 \kappa} \right] =: N^{-1/3} 2a_\kappa =: R_{Newton} .$$

The "general Bohr radius" a_B may be defined by

$$(65) \qquad a_B = \frac{\hbar^2}{Mg^2} \quad \text{with} \quad g^2 = \begin{cases} e^2 & \text{Coulomb } (a_B = a) \\ \kappa M^2 & \text{Newton } (a_B = a_\kappa) . \end{cases}$$

a_κ may be called the "gravitational Bohr radius" in analogy to the hydrogen atom. $a_\kappa \simeq 3.56 \cdot 10^{24}$ cm $\simeq 3.76 \cdot 10^6$ ℓy for neutrons or protons, thus larger than the distance earth to Andromeda galaxy.

"One general Rydberg" may be defined by $E_R = \frac{1}{2\hbar^2} M g^4$ and coincide with 1 Ry for $g^2 = e^2$.

Hence $E_\kappa = \frac{1}{2\hbar^2} M^5 \kappa^2$ may be called "one gravitational Rydberg", "gravitational ionization energy", or "gravitational ground-state energy". According to the weak gravitational coupling we have $E_\kappa \simeq 1.20 \cdot 10^{-69}$ Ry for neutrons or protons, and $E_\kappa^{(e)} \simeq 5.76 \cdot 10^{-86}$ Ry for electrons.

1.2) Nonrelativistic Fermionic Coulomb Systems

The Coulomb forces screen: Every particle interacts only with its nearest neighbour at distance $r = R/N^{1/3}$.
The Coulomb potential energy is

$$(66) \qquad E_{\text{Coulomb}} \simeq -\frac{e^2}{r} \cdot N = -\frac{e^2 N^{4/3}}{R} \quad .$$

With the private room argument and the uncertainty relation:

$$(67) \qquad H \simeq N \frac{p^2}{2M} - \frac{e^2 N}{\hbar} p \; .$$

Minimizing with respect to p leads to the following ground-state energy and mean radius:

$$(68) \qquad H(p_o) \simeq E_N \simeq -N\left[\frac{Me^4}{2\hbar^2}\right] = -N \text{ Ry} \simeq -13.6 \cdot N \text{ eV} \; ,$$

$$(69) \qquad R \simeq N^{1/3} \left[\frac{\hbar^2}{Me^2}\right] = a \cdot N^{1/3} =: R_{\text{Coulomb}} \; .$$

This is in agreement with the exact results (48) and (57).
(Bohr radius $a \simeq 0.53 \cdot 10^{-8}$ cm).

1.3) Nonrelativistic Bosonic Coulomb Systems

No private room argument but screening.

(70) $$H \simeq N\frac{p^2}{2M} - \frac{e^2 p}{\hbar} N^{4/3} , \qquad p_o = N^{1/3}\frac{Me^2}{\hbar} ,$$

(71) $$H(p_o) \simeq E_N \simeq -N^{5/3}\left[\frac{Me^4}{2\hbar^2}\right] = -N^{5/3}\,\mathrm{Ry} ,$$

(72) $$R \simeq N^{-1/3}\left[\frac{\hbar^2}{Me^2}\right] = a\cdot N^{-1/3} .$$

1.4) Nonrelativistic Bosonic Newton Systems

No private room argument and no screening

(73) $$H_N \simeq N\frac{p^2}{2M} - \frac{\kappa M^2}{2\hbar} pN^2 , \qquad p_o = N\cdot\frac{M^3\kappa}{2\hbar} ,$$

(74) $$H_N(p_o) \simeq E_N \simeq -\frac{1}{4}\left[\frac{M^5\kappa^2}{2\hbar^2}\right]\cdot N^3 = -\frac{1}{4}E_\kappa N^3 ,$$

(75) $$R \simeq 2\left[\frac{\hbar^2}{M^3\kappa}\right]\cdot N^{-1} = 2a_\kappa N^{-1} .$$

1.5) Coulomb versus Newton

(76) $$E_{Coulomb} \simeq -\frac{e^2 N^{4/3}}{R} , \quad E_{Newton} \simeq -\frac{\kappa M^2}{2}\frac{N^2}{R} .$$

Both potential energies are of the same magnitude at $N_o = \left(\frac{2e^2}{\kappa M^2}\right)^{3/2} \simeq 10^{54} \simeq$ number of protons/neutrons in Jupiter. N_o depends on the nuclear charges in general. This indicates that gravitational and electrostatic forces are balanced in planets.
Suppose the Pauli principle is valid, then

$$(77) \qquad H_{(N)} \simeq N \frac{p^2}{2M} - \kappa \frac{M^2}{2\hbar} p N^{5/3} - N \frac{e^2 p}{\hbar} ,$$

$$(78) \qquad H_{(N)}(p_o) \simeq E_N \simeq -N \left[\frac{e^4 M}{2\hbar}\right] \left(1 + (\frac{N}{N_o})^{2/3}\right)^2 ,$$

$$(79) \qquad R \simeq R_{Coulomb} \cdot \left((\frac{N}{N_o})^{2/3} + 1 \right)^{-1} .$$

If $N < N_o$ the Coulomb forces dominate, the ground-state energy grows proportional to N, and $R \simeq R_{Coulomb} \sim N^{1/3}$.
If $N > N_o$ the gravitational forces dominate. We observe the $N^{7/3}$-behaviour of the ground-state energy $R \simeq (\frac{N}{N_o})^{-2/3} R_{Coulomb} = R_{Newton} \sim N^{-1/3}$.

1.6) Semirelativistic Fermionic Newton Systems

Let $N_r = \left(\frac{2\hbar c}{\kappa M^2}\right)^{3/2}$. With

$$(80) \qquad H_N \simeq N\sqrt{p^2c^2 + M^2c^4} - \frac{\kappa M^2}{2\hbar} pN^{5/3} ,$$

$$(81) \qquad H_N(p_o) \simeq E_N \simeq N \cdot Mc^2 \left(1 - (\frac{N}{N_r})^{4/3}\right)^{1/2} ,$$

$$(82) \qquad R \simeq \left[\frac{\hbar}{Mc}\right] N^{1/3} \left[1 - (\frac{N}{N_r})^{4/3}\right]^{1/2} \cdot (\frac{N}{N_r})^{-2/3} .$$

The ground-state energy is positive if the particle number is below the critical value N_r. The energy becomes complex in the other case. In the ultrarelativistic limit we observe that for N larger than N_r the Hamiltonian may not be bounded from below:

$$(83) \qquad H_N(p) \simeq Npc - \frac{\kappa M^2}{2\hbar} pN^{5/3} = Npc \; (1 - (\frac{N}{N_r})^{2/3}) .$$

This is minimized by $p \to \infty$ and $H_N(p) \to -\infty$ if $N > N_r$.

The system collapses completely for $N > N_r$. Since semirelativistic gravitating fermions are realized in white dwarf stars, no white dwarf can have a mass larger than $M \cdot N_r$. This is the so-called Chandrasekhar limit, known to astrophysicists since 1931. The first evidence dates back to [23,24 (Chap.XI.2), 25].

Lieb [11] demonstrated that the Heisenberg uncertainty relation cannot provide a finite lower bound to a Hamiltonian with 1/r - singularity. The heuristic arguments deserve a more subtle justification which will be given in the following subsection. It is indeed not Heisenberg's version of the uncertainty relation, i.e.

(84) $$T_\psi(\psi, |x|^2\psi) \geq \frac{3}{4}\frac{\hbar^2}{M} \quad ,$$

which is used in the heuristic arguments, but it is the uncertainty relation (97).

I.ab.2 Energy Inequalities

2.1) Nonrelativistic Cosmic Matter

Purely cosmic matter is described by a fermionic Newton system with Hamiltonian (58) defined on a subspace of the antisymmetrized part of the $\mathcal{L}_2(\mathbb{R}^{3N}, \mathbb{C}^{qN})$, abbreviated as $\mathcal{L}_2^{(-)}$.

Theorem (85): Suppose there exist $m > 0$ and $M > 0$ such that for all $N \in \mathbb{N}$ and for all $i = 1,\ldots,N$: $0 < m \leq M_i \leq M$. The ground-state energy of nonrelativistic purely cosmic matter obeys the following inequality: There exist $A,B > 0$ (independent of N) such that

(86) $$-Aq^{2/3}N(N-1)^{4/3}[\tfrac{1}{2}\,\kappa^2M^5] \leq E_N \leq -Bq^{2/3}N^{1/3}(N-1)^2[\tfrac{1}{2}\,\kappa^2m^5].$$

Proof: Except for slight generalizations we follow [26, Theorem 2].

Lower bound:

(87) $$H_N \geq H_N^\ell \quad ,$$

(88) $$H_N^\ell = \sum_{i=1}^{N} \frac{p_i^2}{2M} - \sum_{1\le i<j\le N} \kappa M^2 |x_i - x_j|^{-1} .$$

From here on the proof follows closely [26] and is therefore only briefly sketched. H_N^ℓ is decomposed according to [27] as a sum of single particle Hamiltonians h_i representing N-1 independent particles j ($\neq$ i) in the field of the fixed ith one:

(89) $$H_N^\ell = \sum_{i=1}^{N} h_i ,$$

(90) $$h_i = \sum_{j\neq i}^{N} \left(\frac{p_j^2}{2(N-1)M} - \kappa \frac{M^2}{2} |x_i - x_j|^{-1} \right) .$$

The lower bound to the fermion ground-state energy is then obtained as usual by filling the levels of the hydrogen spectrum according to the degeneracies.

Upper bound:

For simplicity we confine ourselves to q=1. By the Rayleigh-Ritz variational principle, an upper bound to the fermion ground-state energy is found by the expectation value of H_N with the trial wave function

(91) $$\psi(x_1,\dots,x_N) = \lambda^{\frac{3N}{2}} (N!)^{-\frac{1}{2}} \det[\psi_k(\lambda x_\ell)] .$$

The Slater determinant is composed with the single particle wave functions

(92) $$\psi_k(x) = \psi(x - a_k) ,$$

where the a_k form a lattice with minimal separation $|a_k - a_\ell| \geq 1$ and ψ_k has compact support in a ball of radius 1/2 around a_k. This assures that the exchange term vanishes. Again we use H_N^u instead of H_N where all M_i are replaced by m and $H_N \leq H_N^u$. The rest of the proof follows [26]. □

Remark: For ordinary cosmic matter with

$$(93)\qquad H_N = \sum_{i=1}^{N} \frac{p_i^2}{2M_i} + \sum_{1\leq i<j\leq N} (e_i e_j - \kappa M_i M_j)\,|x_i - x_j|^{-1}$$

one obtains similar results as the heuristic estimates. For fermions we have in (86) the $N^{7/3}$-behaviour and for bosons ($q=N$) the N^3-behaviour in coincidence with the heuristic evaluations. If the $1/r$-potential is regulated at the origin the upper bound in (86) is only affected by a change of the constant B. Thus stability cannot be achieved by cutting of the singularity.

Let $N \geq 2$ and define

$$(94)\qquad \langle P^2\rangle := \frac{1}{N}\sum_{i=1}^{N} \langle p_i^2\rangle \;,$$

$$(95)\qquad \langle |X|^{-1}\rangle := \frac{2}{N(N-1)} \sum_{1\leq i<j\leq N} \langle |x_i - x_j|^{-1}\rangle \;.$$

Corollary (96): The following "uncertainty relations" hold:

a) For $N=1$

$$(97)\qquad T_\psi\,(\psi|\,|x|^{-1}\psi)^{-2} \geq \frac{1}{2}\,\frac{\hbar^2}{M} \;.$$

b) Let $N \geq 2$, then for fermions

$$(98)\qquad \frac{1}{2M}\,\langle P^2\rangle\,\langle |X|^{-1}\rangle^{-2} \geq \frac{1}{4}\,(N-1)^{2/3}\,\frac{\hbar^2}{M} \;.$$

c) Let $N \geq 2$, then for bosons

$$(99)\qquad \frac{1}{2M}\,\langle P^2\rangle\,\langle |X|^{-1}\rangle^{-2} \geq \frac{1}{4}\,N^{-2/3}(N-1)^{2/3}\,\frac{\hbar^2}{M} \;.$$

Proof: The ground-state energy of the one particle Hamiltonian

(100) $$H = \frac{p^2}{2M} - g^2|x|^{-1} = T - g^2|x|^{-1}$$

for arbitrary coupling constant g^2 is known to be

(101) $$E_o = -\frac{Mg^4}{2\hbar^2} .$$

Take expectation values of the inequality $H \geq E_o$ with normalized wave functions ψ. This leads to the inequality

(102) $$T_\psi (\psi|x|^{-1}\psi)^{-2} \geq g^2(\psi|x|^{-1}\psi)^{-1} - \frac{Mg^4}{2\hbar^2}(\psi|x|^{-1}\psi)^{-2} .$$

Since the right-hand side of (102) holds for all g^2 it holds also for the value of g^2 which minimizes the right-hand side of (102). This leads to (97).

To prove (98) and (99) we proceed similarly in starting with the N particle Hamiltonian (58) with $M_i = M$ instead of (100). Applying the bounds (86) for fermions and bosons ($q = N$) completes the proof. ((98) is already known from [26]). □

2.2) Semirelativistic Cosmic Matter

Purely semirelativistic cosmic matter is a fermionic Newton system with the kinetic energy for one particle replaced by

(103) $$H_o = (p^2 + M^2)^{1/2} .$$

Properties of one semirelativistic particle in a central gravitational field:

Theorem (104): Let the domain of definition for a quadratic form A be $\mathcal{Q}(A) = \mathcal{D}(|A|^{1/2}) \subset \mathcal{L}_2(\mathbb{R}^3)$.

a) $H = H_o - g^2|x|^{-1}$ is positive as a quadratic form on $\mathcal{Q}(H_o)$ if $g^2 \leq \frac{2}{\pi}$.

b) H is unbounded from below as a form on $\mathcal{Q}(H_o)$ if $g^2 > \frac{2}{\pi}$.

c) $\mathcal{D}(|x|^{-1}) \supset \mathcal{D}(H_o)$ and H is essentially self-adjoint on $\mathcal{D}(H_o)$ if $g^2 \leq \frac{1}{2}$.

Proof: See [28].

For the semirelativistic N-fermion problem we find:

Theorem (105): Let the Hamiltonian of a system of N gravitating particles be

$$(106)\qquad H_N = \sum_{i=1}^{N} (p_i^2 + M^2)^{1/2} - \sum_{1\leq i<j\leq N} \kappa M^2 |x_i - x_j|^{-1} .$$

a) H_N is a positive form on all of $\mathcal{L}_2(\mathbb{R}^{3N}) \cap \mathcal{Q}(H_N)$ if $N \leq N_{cr,1}$.

b) H_N is not bounded from below on all of $\mathcal{L}_2(\mathbb{R}^{3N}) \cap \mathcal{Q}(H_N)$ if $N > N_{cr,2}$.

c) H_N is not bounded from below on the "antisymmetric subspace" $\mathcal{L}_2^{(-)} \cap \mathcal{Q}(H_N)$ if $N > N_{cr,3}$.

d)

$$N_{cr,1} \simeq \frac{2}{\pi} N_r^{2/3} = \frac{2}{\pi}\left(\frac{2\hbar c}{\kappa M^2}\right),$$

$$N_{cr,2} \simeq 4\pi N_r^{2/3},$$

$$N_{cr,3} \simeq [\sqrt{12}\pi]^{3/2} \cdot N_r \simeq 35.9\, N_r .$$

Proof:

Part a): A Fisher-Ruelle type decomposition of the Hamiltonian (106),

$$(107)\qquad H_N = \frac{1}{N-1} \sum_{i=1}^{N} \left\{ \sum_{j\neq i} [(p_i^2 + M^2)^{1/2} - \frac{\kappa M^2}{2}(N-1)|x_i - x_j|^{-1}] \right\},$$

shows positivity according to the theorem (104a) of Herbst if the coupling constant is small, i.e.

$$(108)\qquad \frac{\kappa M^2}{2}(N-1) \leq \frac{2}{\pi} .$$

Part b): We construct an upper bound to the mean energy which tends to $-\infty$. As variational function we take the symmetric normalized wave function

$$(109)\qquad \Psi(x_1,\ldots,x_N) = \lambda^{3N/2} \bigotimes_{i=1}^{N} \psi(\lambda x_i) ,$$

with

(110) $\psi(x) \sim \chi_{[0,\ell]}(|x|)\sin \frac{2\pi}{\ell}|x|$ with compact support in a ball of radius ℓ and normalized.

By Cauchy-Schwarz inequality:

$$(111) \qquad (\Psi,(p^2 + M^2)^{1/2}\Psi) \leq ((\Psi,p^2\Psi) + M^2)^{1/2} .$$

Consequently we have

$$(112) \qquad (\Psi,H_N\Psi) \leq (\lambda^2\alpha + M^2)^{1/2} N - \frac{\lambda}{2} \kappa M^2 N(N-1)\beta ,$$

with

$$(113) \qquad \alpha = \int |\nabla\psi|^2 d^3x = 4\pi^2\ell^{-2} ,$$

$$(114) \qquad \beta = \iint |\psi(x)|^2|\psi(y)|^2|x-y|^{-1}d^3xd^3y \geq (2\ell)^{-1}.$$

Thus

$$(115) \qquad (\Psi,H_N\Psi) \leq \frac{\lambda}{\ell} N\{(4\pi^2 + \lambda^{-2}M^2\ell^2)^{1/2} - \frac{\kappa M^2}{4}(N-1)\}$$

can be chosen arbitrarily negative if λ/ℓ becomes large enough and

$$(116) \qquad \kappa M^2(N-1) > 8\pi .$$

Part c): We construct an upper bound on the subspace $\mathcal{L}_2^{(-)}$. Choose

$$(117) \qquad \Psi^{(-)}(x_1,\ldots,x_N) = (N!)^{-1/2}\lambda^{3N/2}\det[\psi_k(\lambda x_\ell)]; \quad \psi_k(x) = \psi(x-a_k),$$

with the same single particle wave function $\psi(x)$ as in (110). (111) still holds such that

$$(118) \qquad (\Psi^{(-)}, H_N \Psi^{(-)}) \leq N(\lambda^2\alpha + M^2)^{1/2} - \frac{\lambda}{2}\kappa M^2 \sum_{k\neq \ell} \beta_{k\ell} ,$$

with

$$(119) \qquad \beta_{k\ell} = \iint \rho_{k\ell}(x,y)|x-y|^{-1} d^3x d^3y = |a_k - a_\ell|^{-1} ,$$

$$(120) \qquad \rho_{k\ell}(x,y) = (|\psi_k(x)\psi_\ell(y)|^2 - \overline{\psi}_k(x)\psi_k(y)\psi_\ell(x)\overline{\psi}_\ell(y)) .$$

Suppose that the positions a_k form a three dimensional cubic lattice with lattice constant equal to 2ℓ. All the wave functions ψ_k have disjoint support and the exchange term in (119,120) vanishes. There exists a natural number μ, such that

$$(121) \qquad |a_k - a_\ell| \leq \sqrt{3}\ell(\mu-1) \leq \sqrt{3}\ell N^{1/3} .$$

Thus

$$(122) \qquad (\Psi^{(-)}, H_N \Psi^{(-)}) \leq \frac{\lambda}{\ell} \{(4\pi^2 + \lambda^{-2}M^2\ell^2)^{1/2} N - (2\sqrt{3})^{-1} N^{2/3}(N-1)\kappa M^2\},$$

which concludes the proof with the argument used in part b). □

<u>Open problem</u>: Does there exist a $N_{cr,4} \simeq N_r$, such that H_N is still positive on the "antisymmetric subspace" $\mathcal{L}_2^{(-)} \cap \mathcal{Q}(H_N)$ for $N \leq N_{cr,4}$?

I.b. <u>Canonical and Microcanonical Ensembles</u>

We have studied the scaling behaviour of matter at zero temperature, i.e. in the ground-state, in the previous section a. This enables us to formulate the correct scaling of the thermodynamic functions at positive temperatures for which the thermodynamic limit ought to be studied. Sections b. and c. are devoted to the existence of thermodynamics for matter.

I.ba. <u>The Thermodynamic Limit for Ordinary Matter</u>

This subsection is based on the work of Lieb and Lebowitz in [29,30,31]. For reviews on this subject the reader is referred to the articles of Lieb [11] and Thirring [15].

Stability of ordinary matter was proved by finding an extensive lower bound to the Hamiltonian. This bound involves the short distance behaviour, the 1/r-singularity of the potential and proves that collapse or implosion cannot occur in ordinary matter. Now we are interested in the thermodynamic limit for which the long range r^{-1} behaviour of the Coulomb potential is relevant. If the limit exists it shows that the system does not explode. If the charge excess $\Delta Q = \sum_{i=1}^{K} z_i - N$ is too large, i.e., if the ratio $\frac{|\Delta Q|}{V^{2/3}}$ tends to infinite if $V \to \infty$, $\frac{N}{V} \to \rho$, then the thermodynamic limit does indeed not exist. To assume neutrality, or at least the weaker requirement that the ratio $\frac{|\Delta Q|}{V^{2/3}}$ tends to zero is therefore essential for the existence of the thermodynamic limit. In the third case $|\Delta Q| \sim V^{2/3}$ the thermodynamic limit will be shape dependent.

I.ba.1. Microcanonical Ensemble

We suppose from now on that there is only one sort of nuclei: $z_1 = z_2 = \ldots = z_K = Z$. Furthermore the system is assumed to be neutral, i.e. $N = ZK$. Thus we presume that correlations are not long ranged, despite the long range potential r^{-1}. This is expressed by Newton's theorem, if we take balls as domains, which tend to have infinite volume in the thermodynamic limit: the potential outside of a ball filled with charge in a rotationally symmetric way is the same as a potential generated by the total charge placed in the origin. Therefore the electric potential far away will be approximately zero, if the total charge is small, which means the system is approximately neutral. Newton's theorem is the reason for taking balls as domains.

1.1) Estimating the Interaction Energy between Balls

The Hamiltonian for N electrons with mass m and charge -1 and K nuclei with mass M and charges $z_1,\ldots,z_K$ is a quadratic form with Dirichlet boundary conditions at some ball B:

$$(123) \qquad H_N = \sum_{i=1}^{N}\left(\frac{p_i^2}{2m} - \sum_{\ell=1}^{K} z_\ell |x_i - X_\ell|^{-1}\right) + \sum_{1\le i<j\le N} |x_i - x_j|^{-1} +$$

$$+ \sum_{1\le \ell<n\le K} z_\ell z_n |X_\ell - X_n|^{-1} + \sum_{\ell=1}^{K} \frac{P_\ell^2}{2M_\ell} .$$

We assume $z_\ell = Z$ for all ℓ, $N = ZK$, thus $N_t = N+K = N(1+Z^{-1})$. The wave functions ψ in the domain of H_N fulfill at the boundary ∂B: $\psi(x_1,\dots,x_N, X_1,\dots,X_K) = 0$, if some $x_j \in \partial B$ or some $X_i \in \partial B$.

The first step in proving the existence of the thermodynamic limit of the microcanonical energy density for neutral ordinary matter in a ball B, is filling B with smaller disjoint balls, and estimating the interaction energy between these smaller balls. This leads

Theorem (124): Let a ball B contain k disjoint balls B_α : $B \supset \bigcup_{\alpha=1}^{k} B_\alpha$. Let $N_t = \sum_{\alpha=1}^{k} N_\alpha$ with $N_\alpha/(Z+1) \in \mathbb{N}$ and $S = \sum_{\alpha=1}^{k} S_\alpha$. Let $E(S,V,N_t)$ be the microcanonical energy as function of the entropy, volume $V = |B|$ of the ball B, and total particle number. Then

$$(125) \qquad E(S,V,N_t) \le \sum_{\alpha=1}^{k} E_\alpha(S_\alpha, V_\alpha, N_\alpha).$$

Proof: Consider the quantum Coulomb system of ordinary matter in a ball B of volume $V = |B|$. Let $e_1 \le e_2 \le \dots$ be the eigenvalues of the Hamiltonian arranged in increasing order and including multiplicity. If some eigenvalue e is t-times degenerate then e appears in this list t-times repeated according to the number of the eigenfunctions and the index at the e's denotes all eigenfunctions. Let $S \in \mathbb{R}$, and let $n \ge 1$ be the smallest integer $\ge \exp S$. Then the microcanonical energy is defined by

$$(126) \qquad E(S,V,N_t) = \frac{1}{n} \sum_{i=1}^{n} e_i(V,N_t).$$

Now let B contain two disjoint balls B_1 and B_2 with volumes V_1 and V_2, and let B_1 contain N_1 particles and B_2 contain N_2 particles with $N_t = N_1 + N_2$. In order that we can fill neutral atoms in the balls B_1

and B_2 we require $N_1(Z+1)^{-1} \in \mathbb{N}$ and $N_2(Z+1)^{-1} \in \mathbb{N}$ and let $S = S_1 + S_2$, resp. $n = n_1 n_2$ with $n_1 = [\exp S_1]$ or $n_2 = [\exp S_2]$. Let H_1, H_2 be the Hamiltonian H with N_1, N_2 particles and boundary conditions at $\partial B_1, \partial B_2$. We form the system of trial or variational functions

$$(127) \qquad \psi_{ij} = \psi_i^1 \otimes \psi_j^2 ,$$

where ψ_i^1 and ψ_j^2 are the eigenfunctions of H_1 resp. H_2 and have compact support in B_1 resp. B_2. The ψ_i^1 and ψ_j^2 are antisymmetric in the electron variables and symmetric or antisymmetric in the nuclei coordinates according to their statistics. However statistics are not relevant at all; because of the disjoint support of ψ_i^1 and ψ_j^2 there are no contributions to the interaction energy by the exchange term. Now

$$(128) \qquad \{\psi_i^1, e_i^1\} , \qquad i=1,\dots,n_1,$$

$$(129) \qquad \{\psi_j^2, e_j^2\} , \qquad j=1,\dots,n_2$$

are the first n_1 (resp. n_2) eigenfunctions and eigenvalues (including multiplicity) in B_1 (resp. B_2) with $n=n_1 \cdot n_2$, and the variational wave functions are the $n_1 n_2$ functions in B: ψ_{ij}. By the minimax principle:

$$(130) \qquad E(S,V,N_t) \le \frac{1}{n} \sum_{i=1}^{n_1} \sum_{j=1}^{n_2} (\psi_{ij}, H\psi_{ij})$$

$$= E_1(S_1,V_1,N_1) + E_2(S_2,V_2,N_2) + U$$

with

$$(131) \qquad E_1(S_1,V_1,N_1) = \frac{1}{n_1} \sum_{i=1}^{n_1} e_i^1(V_1,N_1),$$

and E_2 analogously, and

$$(132) \qquad U = \frac{1}{n} \sum_{i=1}^{n_1} \sum_{j=1}^{n_2} U_{ij} ,$$

with

$$(133)\qquad U_{ij} = \sum_{\ell=1}^{N_1} \sum_{m=1}^{N_2} e_\ell e_m \int \frac{d^{3N_1}x\, d^{3N_2}y}{|x_\ell - y_m|} \times$$

$$\times\ |\psi_i^1(x_1,\ldots,x_{N_1})|^2 |\psi_j^2(y_1,\ldots,y_{N_2})|^2 \quad .$$

The index i counting the number of eigenfunctions of the Hamiltonian H_1 in the ball B_1 can best be written as a pair (A,m) with the principle quantum numbers A, including the angular momentum L(A) of the irreducible representation of the rotation group, and m denotes the magnetic quantum number. Now, because of the spherical symmetry, the e_i^1 depend only on A and not on m. They are 2L(A)+1 - times degenerate to each A. Let us first consider the case where n_1 is a perfect quantum number, i.e. it indicates a full L-shell. More precisely, n_1 is such that for every A all the levels (A,m) with $-L(A) \leq m \leq L(A)$ appear in $(1,\ldots,n_1)$ if any one (A,m') does. Then in (132) we are faced with a sum $\sum_{m=-L}^{L} U_{(A,m),j}$ in which the integral

$$(134)\qquad I_A(x) = \sum_{m=-L}^{L} \int |\psi^1_{(A,m)}(x,x_2,\ldots,x_{N_1})|^2 d^3x_2 \ldots d^3x_{N_1} ,$$

which is the average charge density in B_1 has to be evaluated. $I_A(x)$ depends only on the distance of x from the center of B_1, and since we have filled only neutral atoms in B_1 we have with Newton's theorem and (133)

$$(135)\qquad \sum_{i=1}^{n_1} U_{ij} = 0 \quad \text{for all j.}$$

If n_1 is not perfect, it lies between two perfect numbers μ and ν, $\mu < n_1 < \nu$. The sum $\sum_{i=1}^{\mu} U_{ij} = 0$ can be ignored and we are left with $\tilde{U} = \sum_{i=\mu+1}^{n_1} \sum_{j=1}^{n_2} U_{ij}$. Since $\sum_{i=\mu+1}^{\nu} \sum_{j=1}^{n_2} U_{ij} = 0$ we can pick out of the

$\nu-\mu$ quantum numbers just $n_1-\mu$ ones which fulfill $\sum_{i=\mu+1}^{n_1} \sum_{j=1}^{n_2} U_{ij} \leq 0$.

This is always possible, because otherwise the sum $\sum_{i=\mu+1}^{\nu} \sum_{j=1}^{n_2} U_{ij}$ would be strictly positive. Now we relabel the eigenfunctions by taking the $\psi^1_{\mu+1},\ldots,\psi^1_{n_1}$ which produce $\sum_{i=\mu+1}^{n_1} \sum_{j=1}^{n_2} U_{ij} \leq 0$ as variational functions. (Also this means no change in the first two terms in (130) because the $e_{\mu+1} = \ldots = e_{\nu-1}$ by degeneracy). This leads $U \leq 0$, and

(136) $$E(S,V,N_t) \leq E_1(S_1,V_1,N_1) + E_2(S_2,V_2,N_2).$$

The case $k > 2$ is a straightforward generalization, causing only notational complications. The reader is referred to the literature quoted. □

Remarks: Disjointness of the B_α's is necessary, however, how good they may fill B does not affect the validity of the inequality (125). Theorem (124) is independent of statistics.

1.2) Filling a Ball with Balls

The question how good we can fill a ball B by smaller balls B_α is answered in

Lemma (137): Let $R_j = (1+p)^j R_o$, $p \in \mathbb{N}$ and $1+p \geq 27$ be the radii of balls of size j and let B_k be a ball of size k. Then

(138) $$B_k \supset \bigcup_{j=1}^{k-1} (\nu_j \text{ disjoint balls of size } j),$$

(139) $$\nu_j = p^{-1}(1+p)^{3(k-j)}\left(\frac{p}{1+p}\right)^{k-j} \in \mathbb{N}$$

Proof: See [30] or [15].

Remarks:

1) This is a purely geometrical theorem. The geometrical meaning of (137) has caused the name "cheese theorem" [11] or "Emmentalersatz" [15] for it.
2) One can fill a big ball nicely by a certain number of smaller balls. The total volume of the smaller balls is

$$\sum_{j=0}^{k-1} R_j^3 \nu_j = ((1+p)^k R_o)^3 (1 - (\frac{p}{1+p})^k). \tag{140}$$

Only the fraction $(\frac{p}{1+p})^k$ is not filled, which falls exponentially to zero as $k \to \infty$.
3) There are more smaller balls than larger ones, but the ratio

$$\frac{R_j^3 \nu_j}{R_k^3} = \frac{1}{p} (\frac{p}{1+p})^{k-j}$$

is bigger for the larger balls.

1.3) The Monotonicity Argument

We are ready for stating

Theorem (141): The thermodynamic limit exists:

$$\varepsilon(\sigma,\rho) = \lim_{B\to\infty} \frac{1}{|B|} E(\sigma|B|,\rho|B|). \tag{142}$$

Proof: We follow [15] and [30]. Let B_k be filled with disjoint balls B_α. Since B_k cannot be completely filled by the union of the B_α we cannot distribute the particles in a way such that all balls show the same density. (In the usual thermodynamic limit proof we fill a cube by 8 smaller cubes of the same density). Let

$$\frac{N_\alpha}{|B_\alpha|} = \rho(p+1) = \rho_o , \qquad \alpha=1,\dots,\nu_o \tag{143}$$

and

$$\frac{N_\alpha}{|B_\alpha|} = \rho , \qquad \alpha > \nu_o , \tag{144}$$

such that all $\rho_1 = \dots = \rho_k = \rho$. The densities of the balls of size

$j \geq 1$ are the same and coincide with the mean density $\rho = \frac{N_t}{|B_k|}$. The density of the balls of size zero is necessarily larger, such that indeed

(145) $$\rho_k = \frac{1}{|B_k|} \sum_{\alpha} N_\alpha = \frac{1}{p} \sum_{j=0}^{k-1} \rho_j \cdot \left(\frac{p}{1+p}\right)^{k-j}$$

$$= \frac{\rho_o}{p} \left(\frac{p}{1+p}\right)^{k} + \frac{\rho}{p} \sum_{j=1}^{k-1} \left(\frac{p}{1+p}\right)^{k-j} = \rho$$

for $k \geq 1$. The entropy densities $\sigma_j = \frac{S_j}{|B_j|}$ are

(146) $$\sigma_o = \sigma(p+1),$$

(147) $$\sigma_1 = \ldots = \sigma_k = \sigma = \frac{1}{p} \sum_{j=0}^{k-1} \sigma_j \left(\frac{p}{1+p}\right)^{k-j} ,$$

with the mean entropy density $\sigma = \frac{S}{|B_k|}$. Let now $\varepsilon_\ell := \varepsilon_\ell(\sigma_\ell, \rho_\ell)$ denote the energy density $\frac{1}{|B_\ell|} E_\ell(\sigma_\ell |B_\ell|, |B_\ell|, \rho_\ell |B_\ell|)$. We know from (125):

(148) $$E_k(S, |B_k|, N_t) \leq \sum_{j=0}^{k-1} \nu_j E_j(S_j, |B_j|, N_j).$$

Therefore

(149) $$\varepsilon_k = [(1+p)^{3k} |B_o|]^{-1} E_k(S, (1+p)^{3k} |B_o|, N_t)$$

$$\leq \frac{1}{p|B_o|} \sum_{j=0}^{k-1} \left(\frac{p}{1+p}\right)^{k-j} (1+p)^{-3j} E_j(S_j, |B_j|, N_j)$$

$$= \frac{1}{p} \sum_{j=0}^{k-1} \left(\frac{p}{1+p}\right)^{k-j} \varepsilon_j ,$$

and there must exist numbers $c_k \leq 0$, such that

(150) $$\varepsilon_k = c_k + \frac{1}{p} \sum_{j=0}^{k-1} \left(\frac{p}{1+p}\right)^{k-j} \varepsilon_j .$$

The recursive relation (150) is a renewal equation which is solved by induction over k to:

$$(151) \qquad \varepsilon_k = c_k + \frac{1}{1+p}\left(\varepsilon_o + \sum_{j=0}^{k-1} c_j\right).$$

The infinite sum $\lim_{k\to\infty} \sum_{j=0}^{k-1} c_j$ converges by the following reasoning: By (151) and $c_j \leq 0$ it follows that $\varepsilon_k - c_k$ is monotone decreasing, but from (48) for electrons (q=2):

$$(152) \qquad \varepsilon_k - c_k \geq -2.081 \cdot \rho \cdot \frac{(1+Z^{2/3})^2}{(1+Z^{-1})} \quad ,$$

and $\varepsilon_k - c_k$ is therefore bounded from below. This equation (152) is the place where Fermi statistics enter at all in the proof. From the convergence of the infinite sum it follows that

$$(153) \qquad \lim_{k\to\infty} c_k = 0$$

and therefore

$$(154) \qquad \lim_{k\to\infty} \varepsilon_k = \varepsilon(\sigma,\rho) = \frac{1}{1+p}\left(\varepsilon_o + \sum_{j=0}^{\infty} c_j\right)$$

exists. □

Remarks:

1) The domain was ball shaped. It can be generalized to less special domains, but with specified conditions on their shape which allow a "close" packing with balls [30,31].
2) For simplicity we have assumed only one kind of nuclei to be present. The proof carries through equally well for a finite number of sorts of nuclei.
3) For other models of real matter we have the following results: For jellium in which the positive nuclei are described by a fixed, uniform charge background the proof of the existence of the thermodynamic limit was successful [32]. It was however

not successful up to now in the solid state model of matter where the nuclei are fixed point charges of infinite mass, distributed in a non-rotationally invariant manner. This forbids the application of Newton's theorem, which was crucial in our above analysis .

1.4) Convexity

The microcanonical energy density $\varepsilon(\sigma,\rho)$ should possess the significant thermodynamic stability property of convexity, which does not follow directly from the definition for finite volume. It allows us to conclude the positivity of specific heat and compressibility.

As necessary ingredients of the proof of the convexity theorem (169) for the energy density we need the following basic facts about convex functions:

Lemma (155): $f: \mathbb{R} \to \mathbb{R}$ is convex if and only if for all $x_1 < x_2$, $x_1, x_2 \in \mathcal{D}(f)$

$$f(\alpha x_2 + (1-\alpha)x_1) \leq \alpha f(x_2) + (1-\alpha)f(x_1)$$

holds for all $0 < \alpha \leq 1/2$ or for all $1 > \alpha \geq 1/2$.

Proof:

First case: The inequality in (155) is true for $0 < \alpha \leq \frac{1}{2}$. We show that it is also true for $1/2 < \alpha < 1$.

Define the following partition of the interval $[x_1, x_2]$ by

$$(156) \qquad x_{1/2}^{(j)} = 1/2(x_2 + x_{1/2}^{(j-1)}) \in \mathcal{D}(f) \ , \qquad j=1,2,\dots,$$

$$x_{1/2}^{(0)} = x_1 \in \mathcal{D}(f) \ ,$$

or equivalently

$$(157) \qquad x_{1/2}^{(j)} = (1 - (1/2)^j)x_2 + (1/2)^j x_1,$$

because of

$$(158) \qquad 1 - (1/2)^j = \sum_{q=1}^{j} (1/2)^q \ .$$

We have

$$\lim_{j\to\infty} x_{1/2}^{(j)} = x_2 \,. \tag{159}$$

Suppose α is larger than 1/2. Then it is always possible to choose a $0 < \beta \leq 1/2$ and a j large enough, such that

$$(1-\alpha) = (1-\beta)(1/2)^j \,, \tag{160}$$

$$\alpha = \beta + (1-\beta)(1-(1/2)^j) \,, \tag{161}$$

i.e.

$$\alpha x_2 + (1-\alpha)x_1 = \beta x_2 + (1-\beta)x_{1/2}^{(j)} \,. \tag{162}$$

Using the fact

$$f(x_{1/2}^{(j)}) \leq 1/2 f(x_2) + 1/2 f(x_{1/2}^{(j-1)}) \,, \tag{163}$$

one applies the convexity inequality successively j times. This concludes the proof.

Second case: Suppose the inequality in (155) holds for $1 > \alpha \geq 1/2$. Then we apply the same argument as in the first case to the partition

$$x_{1/2}^{(j)} = 1/2(x_1 + x_{1/2}^{(j-1)}) \in \mathcal{D}(f) \,, \quad j=1,2,\ldots, \tag{164}$$

$$x_{1/2}^{(0)} = x_2 \in \mathcal{D}(f) \,.$$

□

Lemma (165): Let $f : \mathbb{R} \to \mathbb{R}$. If

(a) $f(1/2x_1 + 1/2x_2) \leq 1/2f(x_1) + 1/2f(x_2)$ for all $x_1, x_2 \in \mathcal{D}(f)$.

(b) f is monotonic.

then f is continuous in the interior of $\mathcal{D}(f)$.

Proof: If the convexity inequality holds for $\alpha = 1/2$, then it holds also for every integer power of 1/2. This is true because of (157) and (163). Without restriction we assume that $x_1 < x_2$ and $f(x_1) \leq f(x_2)$. The convexity for integer powers of (1/2) leads to the following inequality

(166) $$0 \leq f(x_1 + (1/2)^j (x_2 - x_1)) - f(x_1) \leq (1/2)^j (f(x_2) - f(x_1))$$

from which continuity in the interior of $\mathcal{D}(f)$ follows immediately. □

Remark: It follows also from the monotonicity (together with (166), resp. 1/2-convexity) that the continuity is uniform on compact sets.

Lemma (167): $f: \mathbb{R}^n \to \mathbb{R}$ is convex if and only if

(a) $f(1/2x_1 + 1/2x_2) \leq 1/2f(x_1) + 1/2f(x_2)$ for all $x_1, x_2 \in \mathcal{D}(f)$,

(b) f is continuous in the interior of $\mathcal{D}(f)$.

Proof: We follow [33]. To show the nontrivial direction of the proof we form $g(\alpha) = f(\alpha x_1 + (1-\alpha)x_2) - \alpha f(x_1) - (1-\alpha)f(x_2)$. g is bounded on $[0,1]$ (by (167b)) and $g(0) = g(1) = 0$. Let α_o denote the least value where the maximum g_o of g is attained. Suppose there exist α, x_1, x_2, such that f is strictly concave, then $g_o > 0$ and $\alpha_o \in (0,1)$. Choose a $\delta > 0$ with $[\alpha_o-\delta, \alpha_o+\delta] \subset [0,1]$. Let $y_1 = (\alpha_o-\delta)x_1 + (1-\alpha_o+\delta)x_2 \in \mathcal{D}(f)$ and $y_2 = (\alpha_o+\delta)x_1 + (1-\alpha_o-\delta)x_2 \in \mathcal{D}(f)$. Then we observe that $1/2y_1+1/2y_2 = \alpha_o x_1 + (1-\alpha_o)x_2 \in \mathcal{D}(f)$ and $1/2(\alpha_o-\delta)f(x_1) + 1/2(1-\alpha_o+\delta)f(x_2) + 1/2(\alpha_o+\delta)f(x_1) + 1/2(1-\alpha_o-\delta)f(x_2) = \alpha_o f(x_1)+(1-\alpha_o)f(x_2)$. Applying (167a) to the points y_1 and y_2 leads to the contradiction $0 < g_o = g(\alpha_o) \leq 1/2(g(\alpha_o-\delta) + g(\alpha_o+\delta)) < g_o$. □

The "mid-convexity" or "1/2-convexity" of (165a) or (167a) is not sufficient for convexity. We give the following counterexample [15]:

(168) $$f(x) = \begin{cases} x & \text{on the positive rational numbers in } \mathbb{R} \\ 0 & \text{otherwise in } \mathbb{R}. \end{cases}$$

f has the "1/2-convexity" but is neither monotonic, nor continuous, nor convex.

Theorem (169): The map $\varepsilon : \mathbb{R} \times \mathbb{R}_+ \to \mathbb{R}$ $((\sigma,\rho) \to \varepsilon(\sigma,\rho))$ has the following properties:

(170) $$\varepsilon(\sigma,\rho) \geq -c\rho , \quad c = 2.081(1 + Z^{2/3})^2/(1+Z^{-1}),$$

(171) ε is monotonically increasing in σ,

(172) $\rho^{-1}\varepsilon(\sigma,\rho)$ is monotonically increasing in ρ,

(173) ε is convex.

Proof: (170) follows from the stability of matter, expressed in (48) and neutrality. Concerning the properties (171) and (172) we refer the reader to [15] and shall concentrate on convexity (173), where we also mostly follow [15]. We choose p (in (137)) to be odd, then $\nu_j = (1+p)^{2(k-j)} p^{k-j-1}$ is even for all $0 \leq j \leq k-1$. Let half of the balls be filled with densities (σ,ρ) (resp. $\sigma_o = \sigma(1+p)$, $\rho_o = \rho(1+p)$) and the other half with (σ',ρ') (resp. $\sigma_o' = \sigma'(1+p)$, $\rho_o' = \rho'(1+p)$). Then according to (148) and (149) the renewal sum writes

$$\varepsilon_k(\bar\sigma_k,\bar\rho_k) \leq \frac{1}{2p} \sum_{j=0}^{k-1} \left(\frac{p}{1+p}\right)^{k-j} [\varepsilon_j(\sigma_j,\rho_j) + \varepsilon_j(\sigma_j',\rho_j')], \tag{174}$$

with (compare to (147))

$$\bar\sigma_k = \frac{1}{2p} \sum_{j=0}^{k-1} \left(\frac{p}{1+p}\right)^{k-j} (\sigma_j + \sigma_j') , \tag{175}$$

and (compare to (145))

$$\bar\rho_k = \frac{1}{2p} \sum_{j=0}^{k-1} \left(\frac{p}{1+p}\right)^{k-j} (\rho_j + \rho_j') . \tag{176}$$

When k tends to infinity (174-176) implies (with (150,153)):

$$\varepsilon(1/2(\sigma+\sigma'), 1/2(\rho+\rho')) \leq 1/2(\varepsilon(\sigma,\rho) + \varepsilon(\sigma',\rho')) . \tag{177}$$

The positive quantity $\bar\varepsilon = \varepsilon(\sigma,\rho) + c\rho$ therefore has the 1/2-convexity property. $\bar\varepsilon$ is monotonically increasing in both σ and ρ, because of (171) and (172): Let $\rho_1 < \rho_2$ then $\rho_1^{-1}\bar\varepsilon(\rho_1) \leq \rho_2^{-1}\bar\varepsilon(\rho_2)$ and therefore $\bar\varepsilon(\rho_1) \leq \bar\varepsilon(\rho_2)$ by positivity. $\bar\varepsilon$ is separately continuous in σ on $\mathbb{R}$ and in ρ on $(0,\infty)$ according to (165). We infer with monotonicity in both variables from the remark to (165) and from (166) that the continuity in one variable is not merely uniform but does also hold uniformly in the second variable (on compact sets). Therefore $\bar\varepsilon$ is continuous as a function of two variables in $\mathbb{R} \times (0,\infty) \subset \mathbb{R}^2$, the interior of $\mathbb{R} \times \mathbb{R}_+$. By (167) $\bar\varepsilon$ is convex, thus ε is convex. □

I.ba.2. Canonical and Grand Canonical Ensembles

In Theorem (178) we state the corresponding results on the existence of thermodynamics for the other two ensembles. The proofs have been given in some detail for the microcanonical case, they may be dropped therefore in the following.

Theorem (178): The following thermodynamic limits exist:

(179)
$$f(\beta,\rho) = \lim_{\substack{B\to\infty \\ \frac{N_t}{|B|}\to\rho}} -\frac{1}{\beta|B|} \ell n \mathrm{Tr}\, \exp(-\beta H_N)$$

$$= \inf_{\varepsilon>-c\rho} (\varepsilon - \beta^{-1}\sigma(\varepsilon,\rho)).$$

Let

$$p_B'(\beta,\mu_e,\mu_n) := \frac{1}{\beta|B|} \ell n \sum_{\substack{N=0 \\ N=ZK}}^{\infty} \sum_{K=0}^{\infty} e^{\beta\mu_e N+\beta\mu_n K} \mathrm{Tr}\, \exp(-\beta H_N) ,$$

and

$$p_B(\beta,\mu_e,\mu_n) := \frac{1}{\beta|B|} \ell n \sum_{N=0}^{\infty} \sum_{K=0}^{\infty} e^{\beta\mu_e N+\beta\mu_n K} \mathrm{Tr}\, \exp(-\beta H_N) ,$$

then

(180)
$$p(\beta,\mu_e,\mu_n) = \lim_{B\to\infty} p_B(\beta,\mu_e,\mu_n) = \lim_{B\to\infty} p_B'(\beta,\mu_e,\mu_n) =$$

$$= \sup_{\rho_e=Z\rho_n} (\mu_e\rho_e + \mu_n\rho_n - f(\beta,\rho))$$

with $\rho = \rho_e + \rho_n = (1 + 1/Z)\rho_e$.

Proof: See [30].

Remarks:

1) The proof is similar to the microcanonical case. Equivalence of the ensembles does not follow, one must use the convexity.
2) The supremum which is extended over all densities ρ_e (for

electrons) and ρ_n (for nuclei) is attained in the neutral sector.

3) Generalization to several components is achieved by introducing "isovectors" $\vec{\mu}$ and $\vec{\rho}$, $\vec{N}$, and a charge vector $\vec{E}$. Neutrality writes $\vec{N}\cdot\vec{E} = 0$.

I.bb. The Thomas Fermi Limit

For ordinary matter the usual thermodynamic limits exist and the ensembles are equivalent, the microcanonical energy density is a convex function. All these properties are not true for pure and ordinary cosmic matter, i.e. if gravitational attraction is involved. By Peierls inequality

$$F(\beta,V,N) \leq E_N(V), \tag{181}$$

where F denotes the canonical free energy for inverse temperature β of matter consisting of one sort of N particles enclosed in an open Lebesgue-measurable subset of $\mathbb{R}^3$ of volume V, and $E_N(V)$ denotes the ground-state energy if the system is enclosed in the domain of volume V. The canonical density of the free energy does not exist in the thermodynamic limit because of (86) and (181):

$$\lim_{\substack{V\to\infty \\ \frac{N}{V}\to\rho}} \frac{1}{V} F(\beta,V,N) = -\infty \quad \text{if } \rho \neq 0. \tag{182}$$

This indicates an incorrect scaling used in forming the density of the free energy, which we have treated in case of the non-stable gravitational interaction like a stable interaction.

Remark: The upper bound to the ground-state energy E_N in (86) holds also if the form H_N operates on $\mathcal{L}_2^{(-)}(V)$ because of the compact support of the variational function.

In section aa. we have demonstrated that the scaling for ordinary matter (Coulomb systems) turns out to be proportional to N for the energy and proportional to N for the volume, i.e., the system is extensive. This is not true for Newton systems as the inequalities of section ab. suggest, and indeed the usual thermodynamic limit does not exist, and we should study the free energy normalized by the factor

$N^{-7/3}$. From (64) we let the linear dimensions shrink by a factor $N^{-1/3}$. Necessarily the inverse temperature shrinks like $N^{-4/3}$ if we assume the entropy to be extensive. If N tends to infinity this limit might exist and is called the Thomas-Fermi limit.

In this subsection we are concerned with gravitating fermions either with purely Newton interaction or in the form of "ordinary cosmic matter" as a two component system of one sort of charged gravitating fermions interacting with another sort of oppositely charged gravitating fermions via Coulomb forces.
What we could expect from the Thomas-Fermi limit is summed in the

I.bb.1. Heuristic Considerations

We expect that we can describe the limit system by a single particle distribution $\rho(x,q)$ on the one particle phase space $B \times \mathbb{R}^3$, where the coordinates are restricted to some ball B. ρ should be a probability distribution with

$$(183) \qquad \int \rho(x,q) \frac{d^3q}{(2\pi)^3} = \rho(x) \in \mathcal{L}_1(B).$$

Then from (58) with $M_i = M$:

$$(184) \qquad \langle H_N \rangle = N \iint \frac{p^2}{2M} \rho(x,q) - \frac{\kappa M^2}{2} N^2 \iint \rho(x)\rho(x')|y-y'|^{-1}$$

and with the uncertainty relation and the private room argument:

$$(185) \qquad \lim_{N\to\infty} N^{1/3}|y| = |x| , \quad \text{(See (64))},$$

$$(186) \qquad \lim_{N\to\infty} N^{-2/3} p = q , \quad \text{(See (62))}.$$

Indeed

$$(187) \qquad \lim_{N\to\infty} N^{-7/3}\langle H_N \rangle = E \quad \text{"exists" with}$$

$$(188) \qquad E = \iint \frac{q^2}{2M} \rho(x,q) \frac{d^3q}{(2\pi)^3} d^3x - \frac{1}{2}\iint \rho(x)\rho(x')\kappa M^2 |x-x'|^{-1} d^3x d^3x'.$$

The entropy for one particle distributions and Fermi statistics is known to be

(189) $$S = -\iint \rho \,\ln\rho - \iint (1-\rho)\,\ln(1-\rho) \ .$$

We assume the validity of the variational principle

(190) $$pV = \max_{\rho} \ (\mu\iint\rho - E + \beta^{-1}S).$$

Variation with respect to ρ leads to

(191) $$\frac{\delta p}{\delta\rho} = 0 \ .$$

Let $U[\rho](x) := \int \rho(x')\kappa M^2|x-x'|^{-1}d^3x'$ then from (191)

(192) $$\beta(\frac{q^2}{2M} + U[\rho](x) - \mu)\delta\rho = \delta\rho \ \ln(\rho^{-1} - 1) \ .$$

Finally

(193) $$\rho(x,q) = \left(1 + e^{\beta(\frac{q^2}{2M} + U[\rho](x)-\mu)}\right)^{-1} \ .$$

The probability distribution is determined by a self-consistency equation which bears some resemblance to the Thomas-Fermi equation.

I.bb.2. Local Thermodynamic Functions and their Scaling Properties

We perform the canonical transformation

(194) $$x \to N^{\delta}x, \quad p \to N^{-\delta}p \ ,$$

which is unitary implementable, thus the trace remains unchanged. The length of the boundary (container) is rescaled by $N^{-1/3}$. Consequently

(195) $$\text{replace } \ell \text{ by } N^{-\delta-1/3}\ell \ .$$

Finally we normalize the energy to $N^{-7/3}$. With the abbreviation $\gamma := -1/3-\delta$ we write the rescaled microcanonical energy Σ in terms of the ordinary microcanonical energy $\bar{E}$ as

$$\Sigma(N,S,\ell) = N^{2\gamma-5/3}\bar{E}(N,S,N^{\gamma}\ell,N^{-1/3-\gamma}(\kappa,e)) \tag{196}$$

which does not depend on γ.
The temperature becomes N-dependent because of the definition

$$\beta^{-1} = \frac{\partial\Sigma}{\partial s} = N^{2\gamma-5/3}\cdot N\cdot\frac{\partial\frac{1}{N}\bar{E}}{\partial s} = N^{2\gamma-2/3}\beta'^{-1} . \tag{197}$$

This leads to a rescaled free energy Φ in terms of the usual free energy F like

$$\Phi(N,\beta,\ell) = N^{2\gamma-5/3}F(N,N^{2\gamma-2/3}\beta,N^{\gamma}\ell,N^{-1/3-\gamma}(\kappa,e)) . \tag{198}$$

Since these definitions do not depend on the choice of γ we illustrate the case $\gamma = 0$:

$$\Phi(N,\beta,\ell) = -\frac{1}{N\beta}\,\ell n\mathrm{Tr}_{\mathcal{H}_N^{(-)}(\ell)}\exp(-\beta\tilde{H}_N) , \tag{199}$$

$$\tilde{H}_N = N^{-2/3}\sum_{i=1}^{N}\frac{p_i^2}{2M} - N^{-1}\sum_{1\leq i<j\leq N}\kappa M^2|x_i-x_j|^{-1} . \tag{200}$$

We are interested in the following thermodynamic Thomas-Fermi limits for thermodynamic functions:

$$\Sigma(s,\ell) = \lim_{\substack{N\to\infty \\ \frac{S}{N}\to s}}\Sigma(N,S,\ell) , \tag{201}$$

$$\Phi(\beta,\ell) = \lim_{N\to\infty}\Phi(N,\beta,\ell) . \tag{202}$$

Remarks: For $\kappa = 0$ the Thomas-Fermi limit reduces to the usual thermodynamic limit. The ideal gas has the scaling property

$$\text{(203)} \qquad \lambda^{-1} F_o(\lambda N, \beta, \lambda^{1/3} L) = \lambda^{-7/3} F_o(\lambda N, \lambda^{-4/3}\beta, \lambda^{-1/3} L).$$

The thermodynamic Thomas-Fermi limit is a usual thermodynamic limit for the kinetic energy and a mean field type limit for the potential energy.

I.bb.3. Hamiltonians for Cosmic Matter

In the following subsections we present a generalization of the analysis of the Thomas-Fermi limit by Hertel, Narnhofer and Thirring [34]. All the following proofs belonging to section I.bb. rely heavily on the proofs presented in [34]. No particular reference to [34] will therefore be given at these theorems of I.bb. in general. The reader is advised to consult [34] as well. Our generalization of [34] consists in considering ordinary cosmic matter instead of a kind of "positronium matter" of protons and electrons as it was done in [34]. "Positronium matter" does not lead to the formation of nuclei since strong interaction is neglected; but nuclei are part of ordinary cosmic matter. We define "ordinary cosmic matter" as a gravitating system of $N_1 = N$ electrons with mass M_1 and charge $e_1 = -|e|$ as sort $\alpha = 1$ particles and $N_2 = K = \zeta N$ nuclei (with $\zeta > 0$) of mass $M_2 = M \geq M_1$ and charge $e_2 = Z|e|$ as sort $\alpha = 2$ particles. Neutrality means $\zeta = 1/Z$. All particles are confined to move in the box

$$\text{(204)} \qquad \Lambda_L = \{x \in \mathbb{R}^3 / 0 < x_r < L \quad (r = 1,2,3)\} .$$

The Hilbert space is $\mathcal{H}_{N,L}$, the equivalence class of square-integrable, complex-valued functions $\psi(x_{11},\dots,x_{1N},x_{21}\dots,x_{2K})$, which are antisymmetric in the variables $x_{11},\dots,x_{1N}$ and antisymmetric in the variables $x_{21},\dots,x_{2K}$ with support in $\bar{\Lambda}_L \times \bar{\Lambda}_L \times \dots \times \bar{\Lambda}_L \subset \mathbb{R}^{3(N+K)}$.
The form domain is

$$\text{(205)} \qquad \mathcal{G}_{N,L} = \{\psi \in \mathcal{H}_{N,L},\ \psi \quad \partial \mathop{\times}_{i=1}^{N+K} \bar{\Lambda}_L = 0,\ \psi \text{ is absolutely continuous and}$$

$\nabla_{\alpha i_\alpha}\psi\in\mathcal{H}_{N,L}\}$.

$\nabla_{\alpha i_\alpha}$ denotes the gradient with respect to the αi_α-th variable.

The Hamiltonian is a sesqui-linear form $\mathcal{G}_{N,L}\times\mathcal{G}_{N,L}\to\mathbb{C}$ with

(206) $$(\psi,H\phi)=\sum_{\alpha i_\alpha}\frac{1}{2M_\alpha}\int d(x)\,\overline{(\nabla_{\alpha i_\alpha}\psi(x))}\,\nabla_{\alpha i_\alpha}\phi(x)$$
$$+\frac{1}{2}\sum_{\substack{\alpha\ i_\alpha\\ \beta\ k_\beta}}{}^{'}\int d(x)\overline{\psi(x)}\left[\frac{e_\alpha e_\beta-\kappa M_\alpha M_\beta}{|x_{\alpha i_\alpha}-x_{\beta k_\beta}|}\right]\phi(x)\ .$$

The sum $\sum'$ means that terms with $(\alpha i_\alpha)=(\beta k_\beta)$ are to be discarded.

$$x=(x_{11},\dots,x_{1N},x_{21},\dots,x_{2K})\ ,$$

$$d(x)=\prod_{\alpha=1}^{2}\ \prod_{i_\alpha=1}^{N_\alpha}dx_{\alpha i_\alpha}\ ,$$

$$i_1=1,\dots,N_1=N;\quad i_2=1,\dots,N_2=K=\zeta N.$$

The form (206) is Hermitian and bounded from below, i.e., there exists b, such that

(207) $$(\phi,\phi)_H=(\phi,(H+b)\phi)\geq(\phi,\phi)\quad\text{for all }\phi\in\mathcal{G}_{N,L}\ .$$

$(\|\cdot\|_H,\mathcal{G}_{N,L})$ is a Hilbert space and $\mathcal{G}_{N,L}$ is $\|\cdot\|$-dense in $\mathcal{H}_{N,L}$. The Hamiltonian as operator is constructed by Riesz' theorem: For each $\phi\in\mathcal{H}_{N,L}$ there exists $\psi\in\mathcal{G}_{N,L}$ such that

(208) $$(\phi,\chi)=(\psi,\chi)_H\quad\text{for all }\chi\in\mathcal{G}_{N,L}\ .$$

The map $\phi\to\psi$ is linear, injective, Hermitian, bounded with range $\mathcal{D}_{N,L}$ which is dense in $\mathcal{G}_{N,L}$ with the $\|\cdot\|_H$-norm, consequently dense in $\mathcal{H}_{N,L}$

in the usual norm. Therefore the inverse is self-adjoint on $\mathcal{D}_{N,L}$ and $= \hat{H} + b$.

$$(209) \qquad \hat{H} = -\sum_{\alpha i_\alpha} \frac{\Delta_{\alpha i_\alpha}}{2M_\alpha} + \frac{1}{2}\sum_{\substack{\alpha\ i_\alpha \\ \beta\ k_\beta}}' \frac{e_\alpha e_\beta - \kappa M_\alpha M_\beta}{|x_{\alpha i_\alpha} - x_{\beta k_\beta}|} .$$

$\hat{H}$ has a discrete spectrum only with eigenvalues $E_\nu(N,L,e^2,\kappa)$.

$$(210) \qquad \bar{E}(N,\Omega,L,(\kappa,e)) = \Omega^{-1} \inf_{\mathcal{F}_\Omega} \sum_{\phi\in\mathcal{F}_\Omega} (\phi,H\phi) ,$$

$$(211) \qquad F(N,\beta,L,(\kappa,e)) = -\beta^{-1} \ln \sup_{\mathcal{F}} \sum_{\phi\in\mathcal{F}} e^{-\beta(\phi,H\phi)} .$$

$\mathcal{F}_\Omega(\mathcal{F})$ is a set of Ω (finitely many) mutually orthogonal and normalized vectors of $\mathcal{G}_{N,L}$. By continuity and density arguments we can restrict $\mathcal{F}\subset\mathcal{D}_{N,L}$. Therefore

$$(212) \qquad \bar{E}(N,\Omega,L,(\kappa,e)) = \Omega^{-1} \sum_{\nu=1}^{\Omega} E_\nu(N,L,e^2,\kappa) ,$$

$$(213) \qquad F(N,\beta,L,(\kappa,e)) = -\beta^{-1} \ln \sum_{\nu=1}^{\infty} e^{-\beta E_\nu(N,L,e^2,\kappa)} .$$

I.bb.4. Regulating the Potential

We shall construct the limit by a discretization procedure, starting with the approximation of the 1/r-potential by a suitable regular potential, such that the change in the energy is small compared to $N^{7/3}$.

We take the case $\gamma = 0$ in the scaling behaviour, i.e., we keep the volume fixed $= \Lambda_\ell$. Then

$$(214) \qquad \Phi(\beta,\ell,N) = \frac{-1}{N\beta} \ln \mathrm{Tr}_{\mathcal{H}_{N,\ell}}\, e^{-\beta\tilde{H}_N} ,$$

(215) $$\tilde{H}_N = \hat{T} + \hat{V}_o ,$$

(216) $$\hat{T} = -N^{-2/3} \sum_{\alpha i_\alpha} \frac{\Delta_{\alpha i_\alpha}}{2M_\alpha} ,$$

(217) $$\hat{V}_o = \frac{1}{2N} \sum_{\substack{\alpha\ i_\alpha \\ \beta\ k_\beta}}{}' (e_\alpha e_\beta - \kappa M_\alpha M_\beta) |x_{\alpha i_\alpha} - x_{\beta k_\beta}|^{-1} .$$

We regulate $1/r$ by the function $\frac{1}{r}(1-e^{-\mu r})$ for large $\mu > 0$. We start with the following

<u>Definitions:</u>

(218) $$\begin{Bmatrix} \hat{V}_\mu \\ \hat{U}_\mu \end{Bmatrix} := \frac{1}{2} N^{-1} \sum_{\substack{\alpha\ i_\alpha \\ \beta\ k_\beta}}{}' \begin{Bmatrix} e_\alpha e_\beta - \kappa M_\alpha M_\beta \\ Z^2 e^2 + \kappa M^2 \end{Bmatrix} \frac{e^{-\frac{\mu}{\ell}|x_{\alpha i_\alpha} - x_{\beta k_\beta}|}}{|x_{\alpha i_\alpha} - x_{\beta k_\beta}|} .$$

(219) Subdivide the cube Λ_ℓ in $g=h^3$ cubes $\Lambda^a_{\ell/h}$ $(a=1,\ldots,g)$ of equal length ℓ/h, located at $\ell\xi_a$.

(220)

$\Theta_a(x) = 1$ if x is inside of $\bar{\Lambda}^a_{\ell/h}$,

$\Theta_a(x) = 1/2$ if x is inside a surface of $\bar{\Lambda}^a_{\ell/h}$,

$\Theta_a(x) = 1/4$ if x is on an edge of $\bar{\Lambda}^a_{\ell/h}$,

$\Theta_a(x) = 1/8$ if x is a corner of $\bar{\Lambda}^a_{\ell/h}$,

$\Theta_a(x) = 0$ if x is outside of $\bar{\Lambda}^a_{\ell/h}$.

(221) The number operator of particles of species α in the cube

$\Lambda^a_{\ell/h}$ is $$\hat{N}_{\alpha a} = \sum_{i_\alpha=1}^{N_\alpha} \Theta_a(x_{\alpha i_\alpha}) .$$

(222) A μ,g-dependent c-number:

$$W_{\alpha a,\beta b} := \begin{cases} (e_\alpha e_\beta - \kappa M_\alpha M_\beta)\, \dfrac{(1-e^{-\mu|\xi_a-\xi_b|})}{|\xi_a - \xi_b|\ell} & a \neq b \\ \frac{\mu}{\ell}(e_\alpha e_\beta - \kappa M_\alpha M_\beta) & a = b\ . \end{cases}$$

(223) $$\hat{V}_{\mu g} := \frac{1}{2} N^{-1} \sum_{\substack{\alpha a \\ \beta b}} N_{\alpha a} W_{\alpha a,\beta b} N_{\beta b}\ .$$

Lemma: $\hat{T}$, $\hat{T} + \hat{V}_o$, $\hat{T} + \lambda\hat{U}_o$, $\hat{T} + \lambda\hat{U}_\mu$, and $T + V_{\mu g}$ are self-adjoint operators $\mathcal{D}_{N,\ell} \to \mathcal{H}_{N,\ell}$ if $\mu > 0$ and $\lambda \in \mathbb{R}$. They correspond to sesquilinear Hermitian closed forms T, $T + V_o$, $T + \lambda U_o$, $T + \lambda U_\mu$, and $T + V_{\mu g} : \mathcal{G}_{N,\ell} \times \mathcal{G}_{N,\ell} \to \mathbb{C}$, defined in analogy to (206).

Lemma (224): On the subspace $\mathcal{G}_{N,\ell}$ the following inequality holds: For each $\lambda \in \mathbb{R}_+$:

(225) $$0 \leq U_o \leq \lambda^{-1} T + \lambda a_1 N$$

with

(226) $$a_1 = \frac{1}{4}\, (1+\zeta)^{7/3} q^{2/3} M (Z^2 e^2 + \kappa M^2)^2 .$$

Proof: This estimate results from applying Lévy-Leblond's inequality (86) (the lower bound) to the Hamilton form $\lambda^{-1}(T - \lambda U_o)$. The constant 1/4 in (226) follows from the details of the proof to (85). Notice that $\mathcal{G}_{N,\ell}$ is a fermion subspace. □

Remark: This lemma holds for the operators as well.

In what follows we need an estimate for the number of bound states in a Yukawa potential. A lot of results are known on this subject, we just quote Simon's

Lemma (227): Let $H = -\Delta + \lambda V(r)$ and $\lambda > 0$, with

(a) For all λ, H has no eigenvalues of positive energy and the negative energy spectrum is purely discrete of finite multiplicity,

(b) $I(V) = \int_0^\infty dr \cdot r|V(r)| < \infty$,

(c) $L = \sup_r\, [-r^2 V(r)] < \infty$,

then an upper bound for the total number of bound states (counting multiplicity) $N(\lambda V)$ is

$$N(\lambda V) \leq [L^{1/2} I]\lambda^{3/2} + \lambda I. \tag{228}$$

Proof: See [35].

Lemma (229): On $\mathcal{G}_{N,\ell}$ we have:

$$0 \leq U_\mu \leq \mu^{-1/5}(T + a_2 N) + \mu^{-1/15} a_2' N^{8/9} \tag{230}$$

with

$$a_2 = (1+\zeta)^{5/2} M^{3/2} (Z^2 e^2 + \kappa M^2)^{5/2} \ell^{1/2} q^{2/3} , \tag{231}$$

$$a_2' = (1+\zeta)^{7/3} M^{4/3} (Z^2 e^2 + \kappa M^2)^{7/3} \ell^{1/3} q^{2/3} . \tag{232}$$

Proof: The idea of this proof is due to [36] and consists in computing bounds to the fermion ground-state energy in a Yukawa potential, by considering

$$h = -\Delta - 2M_o \mu^{1/5} N^{-1/3} \lambda_o \frac{e^{-\frac{\mu}{\ell}|x|}}{|x|} \tag{233}$$

as one-particle Hamiltonian, because by the Fisher-Ruelle decomposition [27]:

$$T - \mu^{1/5} U_\mu = \sum_{i=1}^{N+K} H_i \tag{234}$$

with

$$H_i = (2M_o)^{-1} N^{-2/3} \sum_{\substack{j=1 \\ j\neq i}}^{N+K} \left(-\Delta_j - 2M_o \mu^{1/5} N^{-1/3} \lambda_o \frac{e^{-\frac{\mu}{\ell}|x_j - x_i|}}{|x_j - x_i|} \right) \tag{235}$$

and

$$M_o = M(N+K-1), \ \lambda_o = \tfrac{1}{2}(Z^2 e^2 + \kappa M^2). \tag{236}$$

The quantities (227b) and (227c) turn out to become

$$(237)\qquad L = \sup_r \left[r\, e^{-\frac{\mu}{\ell} r} \right] = \frac{\ell}{\mu} e^{-1}$$

and

$$(238)\qquad I = \frac{\ell}{\mu}$$

in this Yukawa potential case. Lemma (227) leads therefore to the following upper bound $N_o = N(\lambda V)$ for the number of bound states (counting multiplicity) in our Yukawa case:

$$(239)\qquad N_o \leq \frac{\lambda_o^{3/2}}{e^{1/2}} \left(\frac{2M_o \ell}{\mu^{4/5} N^{1/3}} \right)^{3/2} + \frac{2M_o \ell \lambda_o}{\mu^{4/5} N^{1/3}} \; .$$

All bound states are higher than those for the hydrogen atom, but we can only fill in N_o particles. Taking the hydrogen spectrum

$$(240)\qquad E_n = -n^{-2} M_o^2 \mu^{2/5} N^{-2/3} \lambda_o^2$$

with degree of degeneracy

$$(241)\qquad g_n = q \sum_{\ell=0}^{n-1} (2\ell+1) = qn^2 \; ,$$

we obtain a lower bound to the Yukawa potential fermion ground-state energy by filling ν_o levels of the hydrogen spectrum. Here

$$(242)\qquad \sum_{n=1}^{\nu_o} g_n \leq N_o \; ,$$

and therefore

$$(243)\qquad \nu_o \leq (3N_o)^{1/3} q^{-1/3} \; .$$

The rest of the proof is a matter of straightforward computation. We have used the fact that if the Hamiltonian H^Y with Yukawa potential is bounded by the Coulomb Hamiltonian H^C, i.e., $H^Y \geq H^C$, then all $E_n^Y \geq E_n^C$

by the minimax principle, because $E_1 \neq -\infty$ and H^Y, H^C are self-adjoint operators with discrete eigenvalues of finite multiplicity. □

Remark: Notice that the inequality (228), which was proved for infinite volume in [35] remains true for finite volume.

Lemma (244): For each $\varepsilon > 0$ and $\mu > 32$ there is a $g_o(\varepsilon,\mu)$ such that for all $g \geq g_o(\varepsilon,\mu)$:

$$(245) \qquad \frac{T + V_{\mu g} - cN}{1 + 2\mu^{-1/5}} \leq T + V_o \leq \frac{T + V_{\mu g} + cN}{1 - 2\mu^{-1/5}} ,$$

with

$$(246) \qquad c = \bar{\varepsilon} + N^{-1} \mu a_3 + \mu^{-1/5} a_2 + 4\mu^{-1/5} a_1 + N^{-1/9}\mu^{-1/15} a_2' ,$$

$$(247) \qquad a_3 = \max\{1,\zeta\}\cdot\{\tfrac{1}{2}(Z^2+1)e^2 + \tfrac{\kappa}{2}(M_1^2 + M_2^2)\}\ell^{-1} ,$$

$$(248) \qquad \bar{\varepsilon} = \zeta\varepsilon(Z^2e^2 + \kappa M^2).$$

Proof: The idea of the proof is an approximation of a continuous function, namely

$$(249) \qquad \frac{(1 - e^{-\frac{\mu}{\ell}|x'-x''|})}{|x'-x''|} ,$$

by a step function

$$(250) \qquad \sum_{a\ b}^{g} \Theta_a(x')\Theta_b(x'') \frac{(1 - e^{-\mu|\xi_a-\xi_b|})}{|\xi_a-\xi_b|\ell} ,$$

which can be done uniformly on the compact set $\bar{\Lambda}_\ell$. The rest of the proof follows from the preceding lemmata. For more details consult [34]. □

Remark: c can be made arbitrarily small if first $N \to \infty$, then $g \to \infty$, and then $\mu \to \infty$.

Lemma (251): Let $\Sigma_{\mu g}$, $\Phi_{\mu g}$ denote the thermodynamic functions where V_o is replaced by $V_{\mu g}$. Then

$$(252) \qquad \frac{\Sigma_{\mu g}(N,\Omega,\ell)-c}{1+2\mu^{-1/5}} \leq \Sigma(N,\Omega,\ell) \leq \frac{\Sigma_{\mu g}(N,\Omega,\ell)+c}{1-2\mu^{-1/5}} ,$$

$$(253) \qquad \frac{\Phi_{\mu g}(N, \frac{\beta}{1+2\mu^{-1/5}},\ell) - c}{1+2\mu^{-1/5}} \leq \Phi(N,\beta,\ell) \leq \frac{\Phi_{\mu g}(N, \frac{\beta}{1-2\mu^{-1/5}},\ell) + c}{1-2\mu^{-1/5}} .$$

Proof: Application of (244). □

I.bb.5. Inserting Walls

Inserting impenetrable walls between the cells changes the specific mean (free) energy $\Sigma_{\mu g}(\Phi_{\mu g})$ by a vanishing amount, as $N \to \infty$. The form domain is the following pre-Hilbert space

$$(254) \qquad \mathcal{G}^W_{N,\ell} = \{\psi \in \mathcal{G}_{N,\ell} / x_{\alpha i_\alpha} \notin \bigcup_{a=1}^{g} \Lambda^a_{\ell/h} \to \psi(x_{11},\ldots,x_{2K}) = 0\} ,$$

and the "wall functions" are the restrictions of the Hamiltonians to this domain, e.g.:

$$(255) \qquad \Phi^W_{\mu g} = \Phi_{\mu g}(H^W_{\mu g}) , \quad H^W_{\mu g} = H_{\mu g} \restriction \mathcal{G}^W_{N,\ell} \times \mathcal{G}^W_{N,\ell} .$$

Lemma (256):

$$\Sigma_{\mu g}(N,\Omega,\ell) \leq \Sigma^W_{\mu g}(N,\Omega,\ell) ,$$

$$\Phi_{\mu g}(N,\beta,\ell) \leq \Phi^W_{\mu g}(N,\beta,\ell) .$$

Proof: $\mathcal{G}^W_{N,\ell} \subset \mathcal{G}_{N,\ell}$. □

As next step we obtain an estimate in the opposite direction:

Lemma (257): For $N > 2^6$:

$$\Sigma^W_{\mu g}(N,\Omega,\ell + N^{-1/6}\ell) \leq \Sigma_{\mu g}(N,\Omega,\ell) + d \ ,$$

$$\Phi^W_{\mu g}(N,\beta,\ell + N^{-1/6}\ell) \leq \Phi_{\mu g}(N,\beta,\ell) + d \ ,$$

$$d = \frac{6g^{2/3}}{\ell^2}\left(\frac{1}{M_1} + \frac{\zeta}{M_2}\right)N^{-1/3} + \zeta B N^{-1/6}$$

with

$$B = \frac{1}{2}\sum_{\substack{\alpha a \\ \beta b}} |W_{\alpha a,\beta b}| \leq 4g^2(Z^2e^2 + \kappa M^2)\mu\ell^{-1} \ .$$

Proof: We insert walls of thickness b and scale the length by a factor $(1 + \ell^{-1}2(h-1)b)$ with $4(h-1)b < \ell$. In particular we choose for $N > 2^6$ the thickness to grow like $4(h-1)b = 2N^{-1/6}\ell$. Consider now the discontinuous mapping $R : \Lambda_{\ell'} \to \Lambda_\ell$, where $\ell' = \ell+2(h-1)b$ is the scaled length, defined by

$$(258) \qquad (Rx)_r = \begin{cases} x_r - 2(n-1)b & \text{if } \zeta_{n-1} + b \leq x_r \leq \zeta_n - b \\ 2\zeta_n - 2nb - x_r & \text{if } \zeta_n - b \leq x_r \leq \zeta_n \\ 2\zeta_n - 2(n-1)b - x_r & \text{if } \zeta_n < x_r \leq \zeta_n + b \end{cases}$$

with $r=1,2,3$; $n=0,1,\dots,h$ and $\zeta_n = n\frac{\ell}{h} + (2n-1)b$. Define next the following function $f : [0,\ell+2(h-1)b] \to \mathbb{R}$ by

(259) $f \in C^1$,

(260) $f^2(x) \leq 1$ for all $x \in [0,\ell+2(h-1)b]$,

(261) $f(x) \neq 1$ for $x \in [\zeta_n-2b,\zeta_n+2b]$ and for all $n=1,\dots,h-1$,

(262) $f(\zeta_n) = 0$ for all $n = 1,\dots,h-1$,

(263) $\xi \to f(\zeta_n+\xi)$ is even for $|\xi| \leq 2b$ and for all $n=1,\dots,h-1$,

(264) $f^2(\zeta_n+\xi)+f^2(\zeta_n+2b-\xi) = 1$ for all $0 \leq \xi \leq b$ and for all $n=1,\dots,h-1$,

(265) $(f'(x))^2 \leq b^{-2}$ for all $x \in [0,\ell+2(h-1)b]$.

Such functions exist. See [34]. Define $I : \mathcal{H}_{N,\ell} \to \mathcal{H}_{N,\ell'}$ by

$$(266) \qquad (I\psi)(x_{11},\ldots,x_{2K}) = \prod_{\alpha i_\alpha} \prod_{r=1}^{3} f(x_{\alpha i_\alpha, r})\psi(Rx_{11},\ldots,Rx_{2K}).$$

Then I is an isometry [37]:

$$(267) \qquad (I\phi, I\psi) = (\phi,\psi) ,$$

and

$$(268) \qquad (I\psi, (T'^W + V'^W_{\mu g})I\psi) \leq (\psi, (T + V_{\mu g})\psi)$$
$$+ \{\frac{3}{2b^2}\left(\frac{1}{M_1} + \frac{\zeta}{M_2}\right) N^{1/3} + \ell^{-1} 2(h-1)b\zeta BN\}(\psi,\psi)$$

holds for $\psi \in \mathcal{G}_{N,\ell}$. Here $T'^W + V'^W_{\mu g}$ denotes the form $T^W + V^W_{\mu g}$ defined on $\mathcal{G}_{N,\ell'} \times \mathcal{G}_{N,\ell'}$. Consequently $I\mathcal{G}_{N,\ell} \subset \mathcal{G}^W_{N,\ell'}$. Therefore every set of normalized and mutually orthogonal vectors in the domain of $T + V_{\mu g}$ is transformed by I into an orthonormal set of vectors in the domain of $T'^W + V'^W_{\mu g}$, which concludes the proof. □

I.bb.6. Distributing the Particles

In the thermodynamic limit only one distribution of the particles among the cubes remain. The difference between the thermodynamic functions belonging to this dominating distribution and the "wall functions" Σ^W and Φ^W vanishes in the thermodynamic limit.

Definitions (269): Let j = 1,2,,...; a = 1,2,...g; α = 1,2, then

a) $X = \{\nu/\nu_{\alpha aj} = 0,1;\ \sum_{aj} \nu_{\alpha aj} = N_\alpha\}$,

b) $[X']_{\Omega'} = \{X''/X'' \subset X' \subset X,\ X'' \text{ contains } \Omega' \text{ elements } (\Omega' \in \mathbb{N})\}$,

c) $Y = \{n/n_{\alpha a} \in \mathbb{N},\ \sum_a n_{\alpha a} = N_\alpha\}$,

d) $X_n = \{\nu \in X/\sum_j \nu_{\alpha aj} = n_{\alpha a} \text{ for } n \in Y\}$,

e) $\mathcal{O} = \{\omega/\omega_n \in \mathbb{N}\ (n \in Y),\ \sum_{n \in Y} \omega_n = \Omega\}$.

Remark (270): $\mathcal{O}$ and Y are finite sets with

$$(271)\qquad \sum_{n\in Y} = \binom{N+g-1}{g-1}\binom{\zeta N+g-1}{g-1} .$$

Definitions (272):

$$\Sigma^{M}_{\mu g}(N,\Omega,\ell) = \min_{n\in Y}\ \inf_{X'\in[X_n]_\Omega}\ \frac{1}{N\Omega}\sum_{\nu\in X'} E_\nu ,$$

$$\Phi^{M}_{\mu g}(N,\beta,\ell) = \min_{n\in Y}\ \{-\frac{1}{\beta N}\,\ell n \sum_{\nu\in X_n} e^{-\beta E_\nu}\} .$$

Lemma (273):

$$\Phi^{M}_{\mu g}(N,\beta,\ell) \geq \Phi^{W}_{\mu g}(N,\beta,\ell) \geq \Phi^{M}_{\mu g}(N,\beta,\ell) - \frac{1}{\beta N}\,\ell n \binom{N+g-1}{g-1}\binom{\zeta N+g-1}{g-1} .$$

Proof: To each distribution $\nu\in X$ corresponds precisely one eigenvalue of $\hat{T}^W + \hat{V}^W_{\mu g}$ (which is the operator corresponding to the form $T^W + V^W_{\mu g}$):

$$(274)\qquad E_\nu = N^{-2/3}\sum_{\alpha aj} \nu_{\alpha aj}\frac{\varepsilon_j}{2M_\alpha} + \frac{1}{2}N^{-1}\sum_{\substack{\alpha aj\\ \beta bk}} \nu_{\alpha aj}W_{\alpha a,\beta k}\nu_{\beta bk} .$$

From Fermi statistics $\nu_{\alpha aj}\in\{0,1\}$, and from the definition it follows that

$$(275)\qquad \Phi^{W}_{\mu g} = -\frac{1}{\beta N}\,\ell n \sum_{\nu\in X} e^{-\beta E_\nu} .$$

We conclude

$$(276)\qquad e^{-\beta N\Phi^{W}_{\mu g}} \leq \sum_{n\in Y}\sum_{\nu\in X_n} \leq e^{-\beta E_\nu} \leq e^{-\beta N\Phi^{M}_{\mu g}}\sum_{n\in Y} 1 .$$

Also it is immediate that

$$(277) \qquad \min_{n\in Y} \{-\frac{1}{\beta N} \ln \sum_{\nu\in X_n} e^{-\beta E_\nu}\} = -\frac{1}{\beta N} \max_{n\in Y} \ln \sum_{\nu\in X_n} e^{-\beta E_\nu}$$

$$\geq -\frac{1}{\beta N} \ln \sum_{\nu\in X} e^{-\beta E_\nu} . \qquad \square$$

<u>Lemma</u> (278):

$$\Sigma^W_{\mu g}(N,\Omega,\ell) \leq \Sigma^M_{\mu g}(N,\Omega,\ell) .$$

<u>Proof</u>: Consider

$$(279) \qquad \Sigma^W_{\mu g} = \frac{1}{N\Omega} \inf_{X'\in[X]_\Omega} \sum_{\nu\in X'} E_\nu = \frac{1}{N\Omega} \min_{\omega\in\mathcal{O}} \sum_{n\in Y} \inf_{X''\in[X_n]_{\omega_n}} \sum_{\nu\in X''} E_\nu$$

$$\leq \frac{1}{N\Omega} \min_{n\in Y} \inf_{X''\in[X_n]_\Omega} \sum_{\nu\in X''} E_\nu .$$

The inequality follows if the minimum over $\omega\in\mathcal{O}$ is restricted to those ω where ω_n vanishes for all but one $n\in Y$. $\square$

<u>Lemma</u> (280): There exists $0 \leq \delta \leq N^{-1}$, such that

$$(1-\delta)\Sigma^M_{\mu g}(N,\Omega',\ell) - \frac{\zeta B}{N} \leq \Sigma^W_{\mu g}(N,\Omega,\ell) ,$$

with

$$N\binom{N+g-1}{g-1} \cdot \binom{N+g-1}{g-1}\Omega' = \Omega .$$

<u>Proof</u>: Suppose the minimum over the $\omega\in\mathcal{O}$ in

$$(281) \qquad \Sigma^W_{\mu g} = \frac{1}{N\Omega} \min_{\omega\in\mathcal{O}} \sum_{n\in Y} \inf_{X''\in[X_n]_{\omega_n}} \sum_{\nu\in X''} E_\nu$$

is attained at $\omega^o\in\mathcal{O}$. Let Y_o denote those particle number distributions $m\in Y$ for which

(282) $$\omega_m^o < \Omega' = \Omega(N \sum_{n\in Y})^{-1}$$

holds.

We define $\delta := \sum_{n\in Y_o} \Omega^{-1}\omega_n^o$ and observe that $0 \leq \delta \leq N^{-1}$. Split the sum $\sum_{n\in Y} = \sum_{n\in Y_o} + \sum_{n\in Y\setminus Y_o}$ in (281) into two parts and estimate the first contribution by

(283) $$\frac{1}{N\Omega}\sum_{n\in Y_o} \inf_{X'\in[X_n]_{\omega_n^o}} \sum_{\nu\in X'} E_\nu \geq \sum_{n\in Y_o} \frac{\omega_n^o}{\Omega} \inf_{\nu\in X} \frac{E_\nu}{N} \geq -\frac{\zeta}{N} B.$$

We shall show that the second contribution leads $(1-\delta)\Sigma_{\mu g}^M(N,\Omega',\ell)$ as a lower bound. Since

(284) $$1 = \sum_{n\in Y} \frac{\omega_n^o}{\Omega}$$

and

(285) $$\Omega' \to \inf_{X'\in[X_n]_{\Omega'}} \frac{1}{\Omega'}\sum_{\nu\in X'} E_\nu$$

is increasing for every $n \in Y$ we have

(286) $$(1-\delta)\,\Sigma_{\mu g}^M(N,\Omega',\ell) = (1 - \sum_{n\in Y_o}\frac{\omega_n^o}{\Omega}) \min_{n\in Y} \inf_{X'\in[X_n]_{\Omega'}} \frac{1}{N\Omega'}\sum_{\nu\in X'} E_\nu$$

$$\leq \sum_{n\in Y\setminus Y_o} \frac{\omega_n^o}{\Omega} \inf_{X'\in[X_n]_{\Omega'}} \frac{1}{N\Omega'}\sum_{\nu\in X'} E_\nu ,$$

$$\leq \sum_{n\in Y\setminus Y_o} \frac{\omega_n^o}{\Omega} \inf_{X'\in[X_n]_{\omega_n^o}} \frac{1}{N\omega_n^o}\sum_{\nu\in X'} E_\nu ,$$

which concludes the proof. □

<u>Remark</u>: B does not depend on N and $\lim_{N\to\infty}\frac{1}{N}\ell n\Omega = \lim_{N\to\infty}\frac{1}{N}\ell n\Omega'$. In the thermodynamic limit only one particle distribution contributes.

I.bb.7. The Dominating Configuration and the Thomas-Fermi Equation

We derive explicit expressions for

(287) $$\Sigma_{\mu g}(s,\ell) := \lim_{N\to\infty} \Sigma^M_{\mu g}(N,\Omega_N,\ell_N) ,$$

(288) $$\Phi_{\mu g}(\beta,\ell) := \lim_{N\to\infty} \Phi^M_{\mu g}(N,\beta_N,\ell_N) ,$$

where Ω_N, β_N and ℓ_N are sequences such that $N^{-1}\ell n\Omega_N$ converges to $s > 0$, β_N converges to $\beta > 0$ and ℓ_N to $\ell > 0$. The main problem treated in this subsection is to interchange the limit $N\to\infty$ with the minimum by which the $\Sigma^M_{\mu g}$ and $\Phi^M_{\mu g}$ are defined.

Definitions (289):

a) $$Z_N = \{z/Nz_{\alpha a} \in \mathbb{N} \, , \, \sum_a z_{1a} = 1 \, , \, \sum_a z_{2a} = \zeta\} ,$$

b) $$Z = \overline{\bigcup_{N\in\mathbb{N}} Z_N} = \{z/z_{\alpha a} \geq 0 \, , \, \sum_a z_{1a} = 1 \, , \, \sum_a z_{2a} = \zeta\} ,$$

c) $$\varepsilon_g(N,\Omega,\ell,z) = \inf_{X' \in [X_n]_\Omega} \frac{1}{N\Omega} \sum_{\nu\in X_n} N^{-2/3} \sum_{\alpha aj} \frac{\nu_{\alpha aj}\varepsilon_j}{2M_\alpha} ,$$

d) $$\Phi_g(N,\beta,\ell,z) = -\frac{1}{\beta N} \ell n \sum_{\nu\in X_n} \exp\{-\beta N^{-2/3} \sum_{\alpha aj} \frac{\nu_{\alpha aj}\varepsilon_j}{2M_\alpha}\} ,$$

e) $$u_{\mu g}(z) = \frac{1}{2} \sum_{\substack{\alpha a\\ \beta b}} z_{\alpha a} W_{\alpha a,\beta b} z_{\beta b} .$$

Lemma (290): The following thermodynamic limit exists:

$$\varepsilon_g(s,\ell,z) = \lim_{N\to\infty} \varepsilon_g(N,\Omega_N,\ell_N,z_N)$$

$$= q \frac{\ell^3}{g} \sum_{\alpha a} \int \frac{d^3p}{(2\pi)^3} \frac{p^2}{2M_\alpha} \left(1 + e^{\beta \frac{p^2}{2M_\alpha} - \beta\mu_{\alpha a}}\right)^{-1}$$

with

$$z_{\alpha a} = q \frac{\ell^3}{g} \int \frac{d^3p}{(2\pi)^3} \left(1 + e^{\beta \frac{p^2}{2M_\alpha} - \beta\mu_{\alpha a}}\right)^{-1} ,$$

and

$$s = \sum_{\alpha a} \{-\beta z_{\alpha a}\mu_{\alpha a} + \beta \frac{\ell^3}{g} \int q \frac{d^3p}{(2\pi)^3} \frac{p^2}{2M_\alpha} \left(1 + e^{\beta \frac{p^2}{2M_\alpha} - \beta\mu_{\alpha a}}\right)^{-1}$$

$$+ \frac{\ell^3}{g} \int q \frac{d^3p}{(2\pi)^3} \ln \left(1 + e^{-\beta \frac{p^2}{2M_\alpha} + \beta\mu_{\alpha a}}\right)\} .$$

Proof: (290) follows from the Legendre transform of the grand canonical ensemble expression. See [38]. □

Lemma (291): The following thermodynamic limit exists:

$$\Phi_g(\beta,\ell,z) = \lim_{N\to\infty} \Phi_g(N,\beta_N,\ell_N,z_N)$$

$$= \sum_{\alpha a} \{z_{\alpha a}\mu_{\alpha a} - \frac{1}{\beta}\frac{\ell^3}{g} \int q \frac{d^3p}{(2\pi)^3} \ln \left(1 + e^{-\beta \frac{p^2}{2M_\alpha} + \beta\mu_{\alpha a}}\right)\}$$

with

$$z_{\alpha a} = \frac{\ell^3}{g} \int q \frac{d^3p}{(2\pi)^3} \left(1 + e^{\beta \frac{p^2}{2M_\alpha} - \beta\mu_{\alpha a}}\right)^{-1} .$$

Proof: (291) follows from the Legendre transform of the grand canonical ensemble expression. See [38]. □

Proposition (292): The following thermodynamic limits exist:

$$\Sigma_{\mu g}(s,\ell) = \lim_{N\to\infty} \min_{z \in Z_N} \{u_{\mu g}(z) + \varepsilon_g(N,\Omega_N,\ell_N,z)\}$$

$$= \inf_{z \in Z} \{u_{\mu g}(z) + \varepsilon_g(s,\ell,z)\} ,$$

and

$$\Phi_{\mu g}(\beta,\ell) = \lim_{N\to\infty} \min_{z \in Z_N} \{u_{\mu g}(z) + \Phi_g(N,\beta_N,\ell_N,z)\}$$

$$= \inf_{z \in Z} \{u_{\mu g}(z) + \Phi_g(\beta,\ell,z)\} .$$

Proof: For each $z \in Z_N$ we denote by $\chi_N(z)$ either $\{u_{\mu g}(z)+\varepsilon_g(N,\Omega_N,\ell_N,z)\}$ or $\{u_{\mu g}(z) + \Phi_g(N,\beta_N,\ell_N,z)\}$. From the continuity of $z \to u_{\mu g}(z)$ and from (290,291) it follows that $\chi_N(z_N)$ converges to $\chi(z)$ for any sequence $z_N \in Z_N$ which converges to $z \in Z$. By inspection of the explicit expressions (289e),(290) and (291) we observe that $z \to \chi(z)$ is continuous and consequently $\inf_{z \in Z} \chi(z)$ is attained by some $\bar{z}$ because of the compactness of Z. Let $z_N \in Z_n$ be a sequence converging to $\bar{z}$ then

$$(293) \qquad \inf_{z \in Z} \chi(z) = \lim_{N\to\infty} \chi_N(z_N) \geq \limsup_{N \to \infty} \min_{z \in Z_N} \chi_N(z).$$

Define a sequence $\bar{z}_N \in Z_N$ by the requirement $\chi_N(\bar{z}_N) = \min_{z \in Z_N} \chi_N(z)$. By compactness of Z there exists an ascending sequence N_i of integers such that $\bar{z}_{N_i}$ converges for $i\to\infty$. Hence,

$$\lim_{i\to\infty} \chi_{N_i}(\bar{z}_{N_i}) = \chi(\lim_{i\to\infty} \bar{z}_{N_i}) \geq \inf_{z \in Z} \chi(z) ,$$

and consequently

$$(294) \qquad \liminf_{N \to \infty} \min_{z \in Z_N} \chi_N(z) \geq \inf_{z \in Z} \chi(z).$$

From (293) and (294) we conclude

$$(295) \qquad \lim_{N\to\infty} \min_{z \in Z_N} \chi_N(z) = \min_{z \in Z} \lim_{N\to\infty} \chi_N(z) ,$$

which proves the proposition. □

To obtain the infimum we perform the variational derivative

$$(296) \qquad \beta^{-1}\delta s = \delta\varepsilon_g - \sum_{\alpha a} \mu_{\alpha a}\, \delta z_{\alpha a} .$$

Variation of Σ at constant s leads

$$(297) \qquad 0 = \sum_{\alpha a} \{\sum_{\beta b} W_{\alpha a,\beta b}\bar{z}_{\beta b} + \bar{\mu}_{\alpha a}\}\delta z_{\alpha a} ,$$

$$(298) \qquad \sum_{a} \delta z_{\alpha a} = 0 ,$$

with the solution

$$(299) \qquad \bar{\mu}_{\alpha a} = \mu_\alpha - \sum_{\beta b} W_{\alpha a,\beta b}\bar{z}_{\beta b} \quad .$$

Therefore the minimizing particle distribution writes

$$(300) \qquad \bar{z}_{\alpha a} = \frac{\ell^3}{g} \int q \frac{d^3p}{(2\pi)^3} \left(1 + e^{\beta[\frac{p^2}{2M_\alpha} + \sum_{\beta b} W_{\alpha a,\beta b}\bar{z}_{\beta b} - \mu_\alpha]}\right)^{-1} ,$$

where μ_1, μ_2 and β are to be determined from

$$(301) \qquad \sum_a \bar{z}_{1a} = 1 \quad , \quad \sum_a \bar{z}_{2a} = \zeta \quad ,$$

$$(302) \qquad s = 2\beta\, u_{\mu g}(\bar{z}) - \beta\mu_1 - \beta\zeta\mu_2 + \\ + \frac{\ell^3}{g} \sum_{\alpha a} \int q \frac{d^3p}{(2\pi)^3} \ln\left(1 + e^{-\beta[\frac{p^2}{2M_\alpha} + \sum_{\beta b} W_{\alpha a,\beta b}\bar{z}_{\beta b} - \mu_\alpha]}\right) + \\ + \beta \frac{\ell^3}{g} \sum_{\alpha a} \int q \frac{d^3p}{(2\pi)^3} \frac{p^2}{2M_\alpha} \left(1 + e^{\beta[\frac{p^2}{2M_\alpha} + \sum_{\beta b} W_{\alpha a,\beta b}\bar{z}_{\beta b} - \mu_\alpha]}\right)^{-1} .$$

If the system (300-302) of self-consistency equations shows more than one solution, then the one has to be chosen, for which the microcanonical energy

$$(303) \qquad u_{\mu g}(\bar{z}) + \frac{\ell^3}{g} \sum_{\alpha a} \int q \frac{d^3p}{(2\pi)^3} \frac{p^2}{2M_\alpha} \left(1 + e^{\beta[\frac{p^2}{2M_\alpha} + \sum_{\beta b} W_{\alpha a,\beta b}\bar{z}_{\beta b} - \mu_\alpha]}\right)^{-1}$$

takes its minimum. This minimal energy is $\Sigma_{\mu g}(s,\ell)$. In the canonical ensemble, analogously, if (300,301) has more than one solution, the one has to be chosen, for which the canonical free energy

(304) $$-u_{\mu g}(\bar{z}) + \mu_1 + \zeta\mu_2 - \frac{1}{\beta}\frac{\ell^3}{g}\sum_{\alpha a}\int q \frac{d^3p}{(2\pi)^3} \ell n\left(1 + e^{-\beta[\frac{p^2}{2M_\alpha} + \sum_{\beta b} W_{\alpha a,\beta b}\bar{z}_{\beta b} - \mu_\alpha]}\right)$$

takes its minimum. This minimal free energy is $\Phi_{\mu g}(\beta,\ell)$.

These self-consistency equations are the Thomas-Fermi equations for the discretized formulation. As last step we prove in the next subsection the existence of the continuum limit.

Remark (305): The system may have a net charge. Neutrality, i.e. $\zeta=1/Z$ is not required, only $\zeta>0$ is necessary. The origin of the existence of the limits, despite of a possible charge excess, is located in the presence of the container with fixed linear dimension ℓ and due to the Thomas-Fermi scaling. If the system would be infinitely extended in space, and extensive, the surplus particles causing a charge excess would repel each other and move to infinity, leading to a surface charge effect (a capacitance) of the system. In this way the total energy can be minimized: without significant increase of the kinetic energy there is a decrease of the interaction energy due to the increasing mutual separation. If the charge excess is too large, the system would explode, and the thermodynamic functions would diverge in the thermodynamic limit. However limits would exist, which are shape dependent according to the above mentioned surface effect, if the electrostatic interaction energy of the surplus particles of charge excess ΔQ is comparable to the total energy, i.e., $(\Delta Q)^2/R \sim E$. For the infinite Coulomb system we have $E \sim N$ and $R \sim N^{1/3}$ and therefore $\Delta Q \sim N^{2/3}$. This is also the situation with which we are confronted in the Thomas-Fermi case, because the charge excess is $\Delta Q = |\zeta Z-1||e|N$, $E \sim N^{7/3}$, and $R \sim N^{-1/3}$, and the ratio of the energies is constant (independent of N). This matter of fact holds regardless of the scaling used. Let us choose the scaling in which the dimensions of the container are kept fixed, the kinetic energy is scaled like $N^{-2/3}$ and the potential energy is scaled like N^{-1}. The above procedure of placing the surplus particles at the boundary of the fixed box would not decrease the total energy, because it costs more kinetic energy than potential energy is gained, and surface effects should not be expected. According to their mutual repulsion the surplus particles would be uniformly distributed at the surface with mean linear spacing $\Delta x \sim (\sqrt{N})^{-1}$. The kinetic energy would increase with the uncertainty principle like $N^{-2/3} \cdot N \cdot (\sqrt{N})^2 \sim N^{4/3}$, the interaction energy would be reduced by $N^{-1}(\Delta Q)^2/\ell \sim N$. Therefore the kinetic energy wins as $N \to \infty$, and this distribution of the particles is not energetically favourable. This argument is again independent of the particular choice of the scaling. It does not hold for the extensive system and is peculiar for the Thomas-Fermi scaling.

I.bb.8 Continuity of the Thomas-Fermi Equation

All estimates together lead

$$(306) \qquad \Sigma(s,\ell) = \lim_{\mu\to\infty} \lim_{g\to\infty} \Sigma_{\mu g}(s,\ell) \;,$$

$$(307) \qquad \Phi(\beta,\ell) = \lim_{\mu\to\infty} \lim_{g\to\infty} \Phi_{\mu g}(\beta,\ell) \;.$$

When the number g of cells tends to infinity, and the cutoff is removed, the question arises, whether these limits can be carried through the Thomas-Fermi equations (300,301) and can be performed inside of the equations. This requires a continuity property of the Thomas-Fermi equations, in order that the limits of the functionals are the functionals of the limits. This is indeed true and we therefore call the next

Theorem (308) ("Continuity of the Thomas-Fermi Equation"):
The above limits (306) and (307) exist and the thermodynamic functions are

$$(309) \qquad \Sigma(s,\ell) = \min_{\rho_\alpha} E_\rho[\rho_\alpha] \;,$$

$$(310) \qquad \Phi(\beta,\ell) = \min_{\rho_\alpha} F_\rho[\rho_\alpha] \;,$$

with

$$(311) \qquad E_\rho = u + \sum_\alpha \int_{\Lambda_\ell} d^3x \int q \frac{d^3p}{(2\pi)^3} \frac{p^2}{2M_\alpha} \left(1 + e^{\beta\left(\frac{p^2}{2M_\alpha} + W_\alpha(x) - \mu_\alpha\right)} \right)^{-1} ,$$

(312) $$F_\rho = -u+\mu_1+\zeta\mu_2-\sum_\alpha \frac{1}{\beta}\int_{\Lambda_\ell} d^3x \int q\frac{d^3p}{(2\pi)^3}\ln\left(1+e^{-\beta[\frac{p^2}{2M_\alpha}+W_\alpha(x)-\mu_\alpha]}\right),$$

(313) $$u = \frac{1}{2}\sum_\alpha \int_{\Lambda_\ell} d^3x\rho_\alpha(x)W_\alpha(x) ,$$

(314) $$W_\alpha(x) = \sum_\beta \int_{\Lambda_\ell} d^3x' \frac{[e_\alpha e_\beta - \kappa M_\alpha M_\beta]}{|x - x'|}\rho_\beta(x') .$$

The ρ_α are determined by the self-consistency equations

(315) $$\rho_\alpha(x) = q\int \frac{d^3p}{(2\pi)^3}\left(1 + e^{\beta[\frac{p^2}{2M_\alpha} + W_\alpha(x) - \mu_\alpha]}\right)^{-1}$$

(316) $$\int_{\Lambda_\ell} \rho_1(x)d^3x = 1 ,$$

(317) $$\int_{\Lambda_\ell} \rho_2(x)d^3x = \zeta .$$

The specific entropy is

(318) $$s = \beta(\Sigma - \Phi).$$

Proof: See [34] and especially the Appendix of [39].

Definition (319): ρ_α^o is a solution of the temperature dependent Thomas-Fermi equation for the microcanonical (canonical) ensemble, if and only if ρ_α^o fulfills the self-consistency equations (315,316 and 317) and $\Sigma(s,\ell) = E_\rho[\rho_\alpha^o]$ $(\Phi(\beta,\ell) = F_\rho[\rho_\alpha^o])$.

Remark (320): As far as the relation

(321) $$\frac{\partial^2\Sigma(s,\ell)}{\partial s^2} > 0$$

is valid, which means, as far as the Legendre transformation is possible, the thermodynamic functions of the microcanonical and canonical ensembles are Legendre transforms of each other, according to (296) and (318). We shall see that (321) does not always hold, and there is a region where the ensembles are inequivalent.

Corollary (322): For each s, $\beta > 0$, and $\ell > 0$ there exists at least one solution of the temperature dependent Thomas-Fermi equations.

Proof: This statement is a direct consequence of (308), in which we have constructed a solution.

I.bc. Properties of the Temperature Dependent Thomas-Fermi Equations

I.bc.1. Interpretation of the Thomas-Fermi Limit

1.1) The Limit $\beta \to \infty$ (for Purely Cosmic Matter)

The name "temperature dependent Thomas-Fermi equation" is justified by the fact that the temperature dependent Thomas-Fermi equation converges to the ground-state Thomas-Fermi equation in the limit $\beta \to \infty$. ((353) corresponds to (30)). We prove this fact for the one component system of purely cosmic matter, enclosed in an open Lebesgue-measurable set $\Lambda \subset \mathbb{R}^3$ of volume V, and with a scaling chosen such that the density $\rho(x)$ is normalized to some positive number n. The generalization to the ordinary cosmic matter system is straightforward.

Lemma (323): Let $\beta_o > 0$ (and sufficiently large) be fixed. There exists a temperature independent number $g_o(n,V)$, such that if $\rho(x)$ is a solution of the self-consistency equations ((315),(316) and (317)) for the one component system of purely cosmic matter, and normalization n instead of 1, then

$$(324) \qquad |\rho(x)| \leq g_o(n,V) \qquad \text{for all } x \in \Lambda, \text{ and for all } \beta > \beta_o.$$

Proof: We choose units in which $2M=1$ and define

$$(325) \qquad g(\beta,y) = q \int \frac{d^3p}{(2\pi)^3} \left(1 + e^{\beta(p^2+y)}\right)^{-1}$$

for arbitrary $y \in \mathbb{R}$. For negative $\overline{y} \in \mathbb{R}_-$ we observe

$$(326) \qquad g(1,\overline{y}) \geq \frac{1}{12\pi^2} |\overline{y}|^{3/2} \quad .$$

Since the normalization is chosen to be

(327) $$\int_{\Lambda} \rho(x) d^3x = n \ ,$$

we conclude from (315) for all $\beta > 0$ and for positive μ:

(328) $$\frac{n}{V} \geq \frac{1}{12\pi^2} \mu^{3/2}$$

(Notice that $W(x) \leq 0$ in our case). If a $\hat{\mu}$ is defined as function of n and V by

(329) $$\frac{n}{V} = \frac{1}{12\pi^2} \hat{\mu}^{3/2} \quad ,$$

then from (328)

(330) $$\mu \leq \hat{\mu}$$

for all possible μ's and all $\beta > 0$. From [42] we take the following bound: There exist numbers k_1, k_2, k_3, such that for all $y \in \mathbb{R}$

(331) $$g(\beta,y) \leq k_1\beta^{-3/2} + k_2\beta^{-1/2}|y| + k_3|y|^{3/2} \ .$$

(See also Chapter II.aa.). Now choose a β_o sufficiently large, such that

(332) $$g(\beta,y) \leq 1 + \frac{1}{6\pi^2} |y|^{3/2}$$

for all $\beta > \beta_o$. Then with [42]:

(333) $$g(\beta,W-\mu) \leq g_o(n,V)$$

for all $\beta > \beta_o$, with

(334) $$g_o(n,V) = 1 + \frac{1}{6\pi^2} \{c_p\bar{d}_q + \hat{\mu}\}^{3/2} \ ,$$

where we have used

(335) $$\sup_{x \in \Lambda} \left| \int_\Lambda d^3y \, \frac{\kappa M^2}{|x-y|} \rho(y) \right| \leq c_p \bar{d}_q \quad ,$$

with

(336) $$c_p = \kappa M^2 \left\| \frac{1}{|x|} \right\|_p \quad , \quad (p < 3) \; .$$

The $|x-y|^{-1}$ are as functions of x in a uniformly bounded set in $\mathcal{L}_p(\Lambda)$ for $p < 3$. Therefore, because of the convex combination:

(337) $$\|W\|_p \leq nc_p \qquad \text{for } p < 3 \quad .$$

Hence with (331) we have (for $\beta > \beta_o$)

(338) $$g(\beta, W-\mu) \leq k_1 \beta_o^{-3/2} + k_2 \beta_o^{-1/2} \{|W| + \hat{\mu}\} + k_3 \{|W| + \hat{\mu}\}^{3/2} \quad ,$$

and we conclude that the $\rho(x)$ are in a bounded set in every $\mathcal{L}_{2p/3}(\Lambda)$. Therefore there exists a number $\bar{d}_q$, which is independent of β, such that

(339) $$\|\rho\|_q \leq \bar{d}_q = \bar{d}_q(n,V)$$

for all $q < 2$. Now (335) follows by Hölder inequality if p and q are chosen in a way, such that

(340) $$2 < p < 3 \;, \quad 3/2 < q < 2 \;, \quad p^{-1} + q^{-1} = 1. \qquad \square$$

Lemma (341): For purely cosmic matter the functional

$$W[\rho](x) = -\kappa M^2 \int_\Lambda \frac{1}{|x-y|} \rho(y) d^3y$$

is continuous in the $\mathcal{L}_\infty$-topology, and

(342) $$|W[\rho_1](x) - W[\rho_2](x)| \leq c_1(n,V) \|\rho_1 - \rho_2\|_\infty .$$

For $\beta > \beta_o$ all density profiles ρ are elements of the set

(343) $$\Omega_{n,V} := \{\rho \in \mathcal{L}_\infty(\Lambda) \cap \mathcal{L}_1(\Lambda) / \rho \geq 0, \int_\Lambda \rho(x) d^3x = n, \rho \leq g_o(n,V)\}.$$

$\Omega_{n,V}$ is closed in the $\mathcal{L}_\infty$-topology.

Lemma (344): For purely cosmic matter the chemical potential μ shows the following continuity property on $\Omega_{n,V}$: If $\underset{\beta \to \infty}{\text{s-lim}}\, \rho_\beta = \rho_\infty \in \Omega_{n,V}$ exists, then the limit

(345) $$\lim_{\beta\to\infty} \mu[\rho_\beta] = \mu_\infty$$

exists, and is a functional of ρ_∞.

Proof: From the self-consistency equation (315) and from (327) we obtain the following bound

(346) $$n \leq \beta_o^{-3/2} V\{q \int \frac{d^3\overline{p}}{(2\pi)^3} e^{-\overline{p}^2}\} e^{-\beta(-c_p\overline{d}_q - \mu[\rho_\beta])} ,$$

where the indices p and q have to be chosen as in (340). This yields

(347) $$\liminf_{\beta \to \infty} \mu[\rho_\beta] \geq -c_p\overline{d}_q .$$

Since

(348) $$\lim_{\beta\to\infty} g(\beta, y_\beta) = q \int \frac{d^3p}{(2\pi)^3} \Theta(p^2 + y_\infty)$$

converges to the integrated step function if $\lim_{\beta\to\infty} y_\beta = y_\infty$ for some sequence y_β, and the right-hand side of (348) is monotone in y_∞, we conclude by choosing $y_\beta = W[\rho_\beta](x) - \mu[\rho_\beta]$ and considering in (315) the $\limsup_{\beta \to \infty}$ and the $\liminf_{\beta \to \infty}$, that, by monotonicity,

(349) $$\liminf_{\beta \to \infty} \mu[\rho_\beta] = \limsup_{\beta \to \infty} \mu[\rho_\beta] = \mu_\infty .$$ □

The results of the last three lemmata together prove the following

<u>Proposition</u> (350): There exists a $\beta_o > 0$, such that if ρ_β is a solution of the temperature dependent Thomas-Fermi equation, then $\rho_\beta \in \Omega_{n,V}$ for $\beta > \beta_o$.

If in the $\mathcal{L}_\infty$-topology $\text{s-lim}_{\beta \to \infty} \rho_\beta = \rho_\infty \in \Omega_{n,V}$ exists, then the following limits exist

$$(351) \qquad \lim_{\beta\to\infty} \mu[\rho_\beta] = \mu_\infty[\rho_\infty] \quad ,$$

$$(352) \qquad \lim_{\beta\to\infty} W[\rho_\beta](x) = W[\rho_\infty](x) \qquad \text{for each } x \in \Lambda \ ,$$

and ρ_∞ is a solution of the ground-state Thomas-Fermi equation

$$(353) \qquad \rho_\infty(x) = \frac{q}{6\pi^2}\left(\max\{\mu_\infty[\rho_\infty] - W[\rho_\infty](x), 0\}\right)^{3/2} .$$

<u>Remark</u> (354): By weak compactness of $\Omega_{n,V}$ and by uniqueness of the solutions of the ground-state Thomas-Fermi equation there is at least weak convergence of ρ_β to ρ_∞. Proposition (350) remains true if the presupposition of strong convergence of ρ_β to ρ_∞ with respect to the $\mathcal{L}_\infty$-topology is replaced by weak convergence, as can be seen by obvious modifications of the proofs.

1.2) <u>Mean Field Theory</u>

The Thomas-Fermi limit has the character of a mean field type limit. Since

$$(355) \qquad \Phi(N,\beta,\ell) = N^{2\gamma-5/3} F(N, N^{2\gamma-2/3}\beta, N^\gamma \ell, N^{-1/3-\gamma}(\kappa,e))$$

does not depend on the choice of γ we take $\gamma = 1/3$. Then

$$(356) \qquad V_N = N\ell^3 \to \infty \ , \quad \frac{N}{V_N} = \ell^{-3} = \rho \text{ fixed} \ ,$$

$$(357) \qquad \beta_N = \beta \text{ fixed} \ ,$$

(358) $\kappa_N = N^{-2/3}\kappa$.

The Thomas-Fermi limit appears as an infinite volume limit.

1.3) Scaling Properties of $\Phi(\beta,\ell)$

One observes the following scaling property of the free energy $\Phi(\beta,\ell) =: \Phi_{W_{(1)}}(1,\beta,\ell,\kappa)$ (from (198) and (202)):

(359) $$\Phi_{W_{(1)}}(1,\beta,\ell,\kappa) = \lambda^{2\gamma-5/3}\Phi_{W_{(\lambda)}}(\lambda,\lambda^{2\gamma-2/3}\beta,\lambda^{\gamma}\ell,\lambda^{-1/3-\gamma}\kappa) .$$

The potential $W_{(1)}$ generates a density $\rho_{(1)}$ via the self-consistency equation (315) with

(360) $$\int_{\Lambda_\ell} \rho_{(1)}(x)d^3x = 1$$

and $W_{(\lambda)}$ is

(361) $$\lambda^{2\gamma-2/3}W_{(\lambda)}(\lambda^{\gamma}x) = W_{(1)}(x) ,$$

which generates the density $\rho_{(\lambda)}$ with

(362) $$\int_{\Lambda_{(\lambda^{\gamma})\ell}} \rho_{(\lambda)}(x)d^3x = \lambda .$$

Let $\lambda = N \in \mathbb{N}$ be some particle number, then for each $\varepsilon > 0$ there is an N sufficiently large such that

(363) $$|F(N,\beta_N,\ell_N,\kappa_N) - \Phi_{W_{(N)}}(N,\beta_N,\ell_N,\kappa_N)| < \varepsilon.$$

The parameters β, ℓ, and κ have no physical significance. The physical inverse temperature of the system is

(364) $\beta_N = N^{2\gamma-2/3}\beta$,

the physical linear dimension of the system is

(365) $\ell_N = N^{\gamma}\ell$,

and the gravitational coupling constant is

(366) $\kappa_N = N^{-1/3-\gamma}\kappa$,

after N and γ have been chosen.

For neutral ordinary matter ($\kappa = 0$):

(367) $|\rho^{-1}f(\beta_N,\rho) - \Phi_{W_{(N)}}(N,\beta_N,\ell_N,\kappa = 0,\ e_N^2)| < \varepsilon$,

(368) $|\rho - N^{1-3\gamma}\ell^{-3}| < \varepsilon$,

with $e_N^2 = N^{-1/3-\gamma}e^2$ being the square of the unit charge. e^2 is an unphysical scaling parameter.

I.bc.2 The Pressure for a Spherical Symmetric Boundary

The pressure in the canonical ensemble is defined by

(369) $p_\rho := -\frac{\partial}{\partial V} F_\rho(n,\beta,\ell)$, $V = \ell^3$ = volume ,

where $F_\rho(n,\beta,\ell)$ is given by the scaled ($\lambda = n$) expression (312).

Lemma (370) (Virial theorem): For each solution ρ of (315,316,317)

(371) $3\ell^3 p_\rho = 2\ E_{\rho,kin} + E_{\rho,pot}$

with

$$E_{\rho,kin} = \sum_{\alpha} \int_{\Lambda_\ell} d^3x \int q \frac{d^3p}{(2\pi)^3} \frac{p^2}{2M_\alpha} \left(1 + e^{\beta(\frac{p^2}{2M_\alpha} + W_\alpha(x) - \mu_\alpha)}\right)^{-1} \tag{372}$$

and

$$E_{\rho,pot} = \sum_{\alpha} \int_{\Lambda_\ell} d^3x \, \rho_\alpha(x) W_\alpha(x) \ . \tag{373}$$

Proof: The proof follows [39,15]. From the scaled expression (312) one derives

$$\frac{\partial F_\rho}{\partial n} = \mu_1 + \zeta\mu_2 \ , \tag{374}$$

$$\beta \frac{\partial F_\rho}{\partial \beta} = E_\rho - F_\rho \ . \tag{375}$$

With $\frac{p^2}{2M_\alpha} = \varepsilon_\alpha$, partial integration leads (for each $y_\alpha \in \mathbb{R}$):

$$\int_0^\infty d\varepsilon_\alpha \sqrt{\varepsilon_\alpha} \, \ell n \left(1 + e^{-\beta(\varepsilon_\alpha + y_\alpha)}\right) = \frac{2}{3} \beta \int_0^\infty d\varepsilon_\alpha \cdot \varepsilon_\alpha^{3/2} \left(1 + e^{\beta(\varepsilon_\alpha + y_\alpha)}\right)^{-1} \ . \tag{376}$$

Consequently

$$F_\rho = n(\mu_1 + \zeta\mu_2) - \frac{2}{3} E_{\rho,kin} - E_{\rho,pot} \ . \tag{377}$$

Dilatation with $\lambda \in \mathbb{R}_+$ yields

$$F_\rho(n,\beta,\ell) = \lambda^{-7/3} F_\rho(\lambda n, \lambda^{-4/3}\beta, \lambda^{-1/3}\ell) . \tag{378}$$

From (378) we conclude after differentiation with respect to λ

$$0 = -\frac{7}{3} F_\rho + (\mu_1 + \zeta\mu_2)n - \frac{4}{3}(E_\rho - F_\rho) + \ell^3 p_\rho \ , \tag{379}$$

which concludes the proof. □

The local pressure may be defined as

(380) $$p_\rho(x) := 2/3\, E_{\rho,kin}(x) =$$

$$= 2/3 \sum_\alpha \int q \frac{d^3p}{(2\pi)^3} \frac{p^2}{2M_\alpha}\left(1+e^{\beta\left(\frac{p^2}{2M_\alpha} + W_\alpha(x) - \mu_\alpha\right)}\right)^{-1} ,$$

because by partial integration

(381) $$\nabla p_\rho(x) = -\sum_\alpha \rho_\alpha(x)\, \nabla W_\alpha(x)$$

the equation for hydrostatic equilibrium is valid.

Up to now we have not used boundary conditions. We introduce a spherical symmetric boundary by placing the system in a ball-shaped container of radius R. Let $r = |x|$. Then (381) writes

(382) $$\frac{d}{dr} \frac{2}{3} E_{\rho,kin}(r) = -\sum_\alpha \frac{\rho_\alpha(r) M_\alpha(r)}{r^2} ,$$

(383) $$M_\alpha(r) = -\sum_\beta (e_\alpha e_\beta - \kappa M_\alpha M_\beta) \int_0^r dr' 4\pi r'^2 \rho_\beta(r') ,$$

because according to [42] the solutions $\rho(x)$ become rotational invariant and depend only on r. There is a connection between the local pressure and the pressure:

Lemma (384): For spherical symmetric boundary conditions

(385) $$p_\rho = p_\rho(R) .$$

Proof: We follow the proof given in [39,15]. Integrate (382) in the following way

$$\frac{4\pi}{3} \int_0^R dr \cdot r^3 \;.\; (382),$$

then,

$$(386)\qquad V\,\frac{2}{3}\,E_{\rho,kin}(R) - \frac{2}{3}\,E_{\rho,kin} = \frac{4\pi}{3}\int_0^R dr\cdot r^3\cdot\frac{d}{dr}\,\frac{2}{3}\,E_{\rho,kin}(r) =$$

$$= \sum_{\alpha,\beta} \frac{1}{3}\,[e_\alpha e_\beta - \kappa M_\alpha M_\beta]\int_0^R dr\cdot r\cdot 4\pi\rho_\alpha(r)\int_0^r dr'\cdot r'^2\cdot 4\pi\rho_\beta(r')$$

$$= \frac{1}{3}\,E_{\rho,pot}\ . \qquad \Box$$

I.bc.3 Heuristic Considerations on the Possibility of a Phase Transition

We give a heuristic discussion on the possibility of the existence of a gravitational instability, which is due to Thirring [40,15]. The rigorous theory of the gravitational phase transition will be presented in part III.

Choose purely cosmic matter and 2M=1.
We suppose that the free energy is composed of the zero point energy for fermions, the classical free energy for free particles, and the gravitational energy, thus

$$(387)\qquad \frac{F}{N} = \left(\frac{N}{V}\right)^{2/3} - T\ell n\,\frac{V}{N}\,T^{3/2} - \frac{\kappa N}{V^{1/3}}\ .$$

Consequently entropy, energy, and pressure are

$$(388)\qquad \frac{S}{N} = -\frac{1}{N}\,\frac{\partial F}{\partial T} = \ell n\,\frac{V}{N}\,T^{3/2} + 3/2\ ,$$

$$(389)\qquad \frac{E}{N} = \frac{F}{N} + T\,\frac{S}{N} = \left(\frac{N}{V}\right)^{2/3} + \frac{3}{2}\,T - \frac{\kappa N}{V^{1/3}}\ ,$$

$$(390)\qquad p = -\frac{\partial F}{\partial V} = \frac{2}{3}\,\left(\frac{N}{V}\right)^{5/3} + \frac{NT}{V} - \frac{\kappa N^2}{3V^{4/3}} =$$

$$= \frac{E - E_p/2}{\frac{3}{2}\,V}\ ,\quad E_p = -\frac{\kappa N^2}{V^{1/3}}\ .$$

The quantum effects dominate for small V, the classical effects for large V. In the intermediate regime there is the possibility that the pressure p becomes negative and the system condensates to a smaller volume V_o. The relation between the temperature and the energy T=f(E) is linear above the condensation point, i.e. for $E > -\frac{\kappa N^2}{2V^{1/3}}$ and quadratic below the condensation point. Consequently there is a region of negative specific heat. For more details see [15].

I.bc.4. Numerical Solutions of the Thomas-Fermi Equations

4.1) Graph: Entropy versus Total Energy [39, Fig.4],[15, T8, p.141], [15 ,Vol.4, p.238],[40, Fig.6]

a) For R = 100 km there is a region where S(E) is not concave. Microcanonical and canonical ensembles are inequivalent.
b) There is a region where $C_V < 0$. In this region an instability occurs.

4.2) Graph: Total Energy versus Inverse Temperature [39, Fig.3], [15, T8, p.140],[15, Vol.4, p.237],[40, Fig.7]

a) The released energy $4.5 \cdot 10^{52}$ erg is comparable to energies released in supernova explosions.
b) $T_{critical} \simeq 7 \cdot 10^{10}$ K.
c) In the canonical ensemble the region of negative specific heat (for R = 100 km) is bridged by a phase transition. Maxwell's construction applies.

4.3) Graph: Free Energy/Temperature versus Inverse Temperature [39, Fig.2],[15, T8, p. 143],[15, Vol.4, p.239]

a) The free energy is the minimum of the branches.
b) There are three solutions for the self-consistency equation. Two solutions give rise to the same minimal free energy and a non-differentiable behaviour.

4.4) Graph: Mass Distribution versus Radial Distance [39, Fig.5], [15, T8, p.144],[15, Vol.4, p.240],[40, Fig.8]

a) $\rho(r)$ is monotonically decreasing.
b) For $T \simeq 7 \cdot 10^{10}$ K there are two solutions of the Thomas-Fermi equation.

c) The central density reaches 10^{14} - 10^{15} g/cm^3 ≃ nuclear density ≃ density in the interior of neutron stars.

d) After cooling down the created neutron star attains a radius of ≃ 15.9 km.

e) The central density below the transition temperature is more than 10^5 times larger than the density of the surface.

4.5) Graph: Fermi Degeneracy versus Radial Distance [39, Fig.6], [15, T8, p.145],[15, Vol.4, p.241],[40, Fig.9]

a) $$\zeta(r) = \frac{3}{2} \frac{\rho(r)}{\beta E_{\rho,kin}(r)} .$$

b) Complete Fermi degeneracy $\zeta = 0$, Boltzmann statistics $\zeta = 1$.

c) The transition turns a Boltzmann gas into a degenerate Fermi gas with a small atmosphere of classical Boltzmann gas.

4.6) Graphs: Pressure versus Volume [39, Fig.7], and [15, T8, p. 142], [15, Vol.4, p. 236],[40, Fig.5]

a) The negative compressibility is avoided by the phase transition.

b) The dashed line corresponds to an ideal Boltzmann gas.

4.7) Summary

a) The computer solutions indicate a region of negative specific heat in the microcanonical ensemble and a phase transition in the canonical ensemble.

b) The data are fitted to a supernova explosion. The gas phase is the core of a pre-supernova, the condensed phase corresponds to the created neutron star. The transition is an implosion of the stars interior. To treat the explosion of the outer shells, which follows the implosion of the core, is beyond the scope of our model.

Sources: Figures 4.1, 4.2, 4.3, 4.4, 4.5, and the lower figure 4.6 are reproduced from W. Thirring, Vorlesungen über Mathematische Physik T8 [15], Originalmanuskript, with kind permission. The upper figure 4.6 is reproduced from P. Hertel and W. Thirring, in Quanten und Felder [39], Vieweg-Verlag, with kind permission.

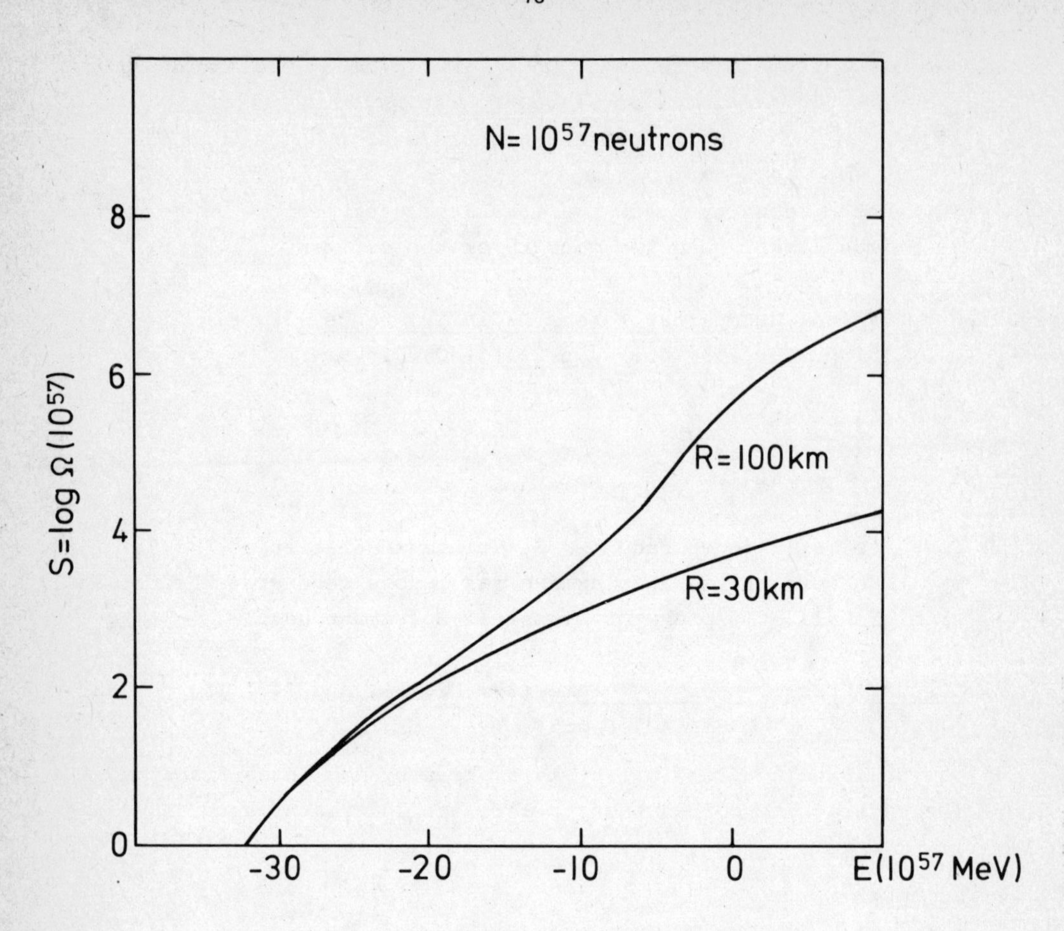

Ad 4.1): Entropy versus Total Energy

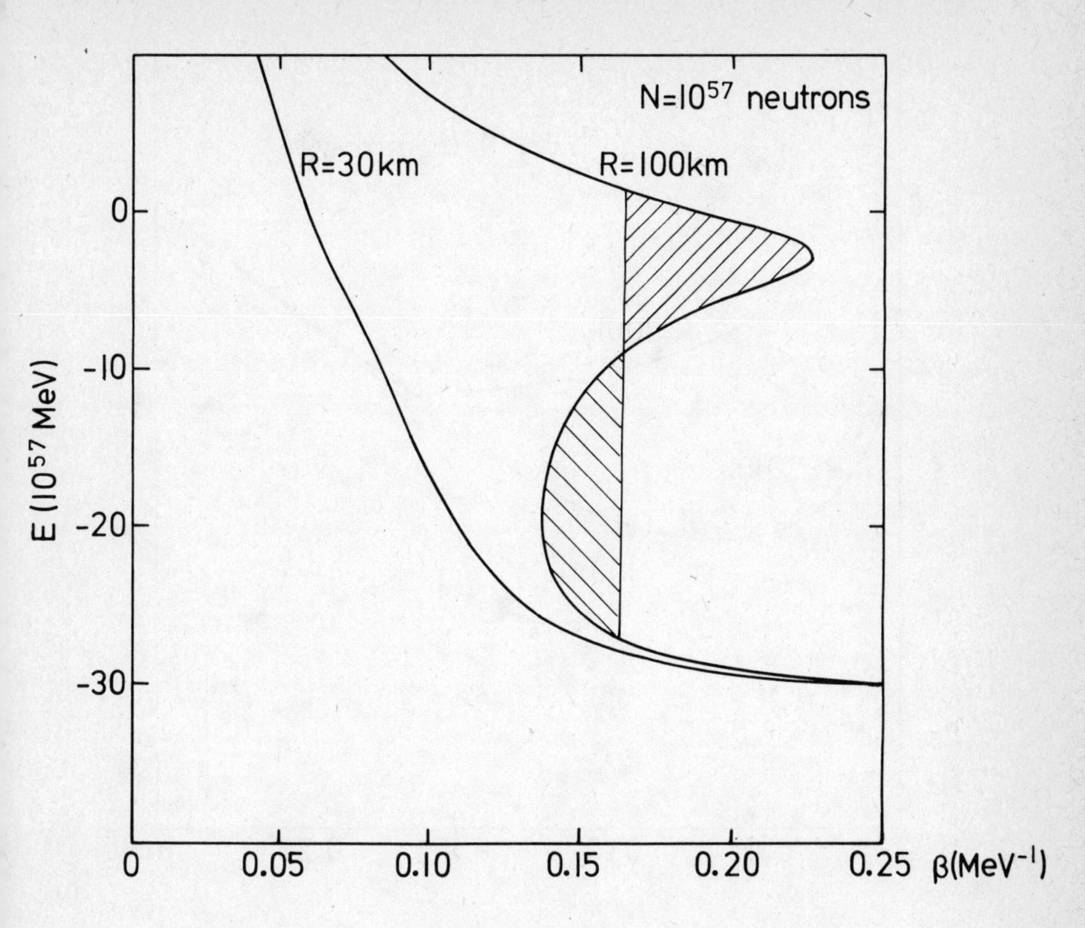

Ad 4.2): Total Energy versus Inverse Temperature

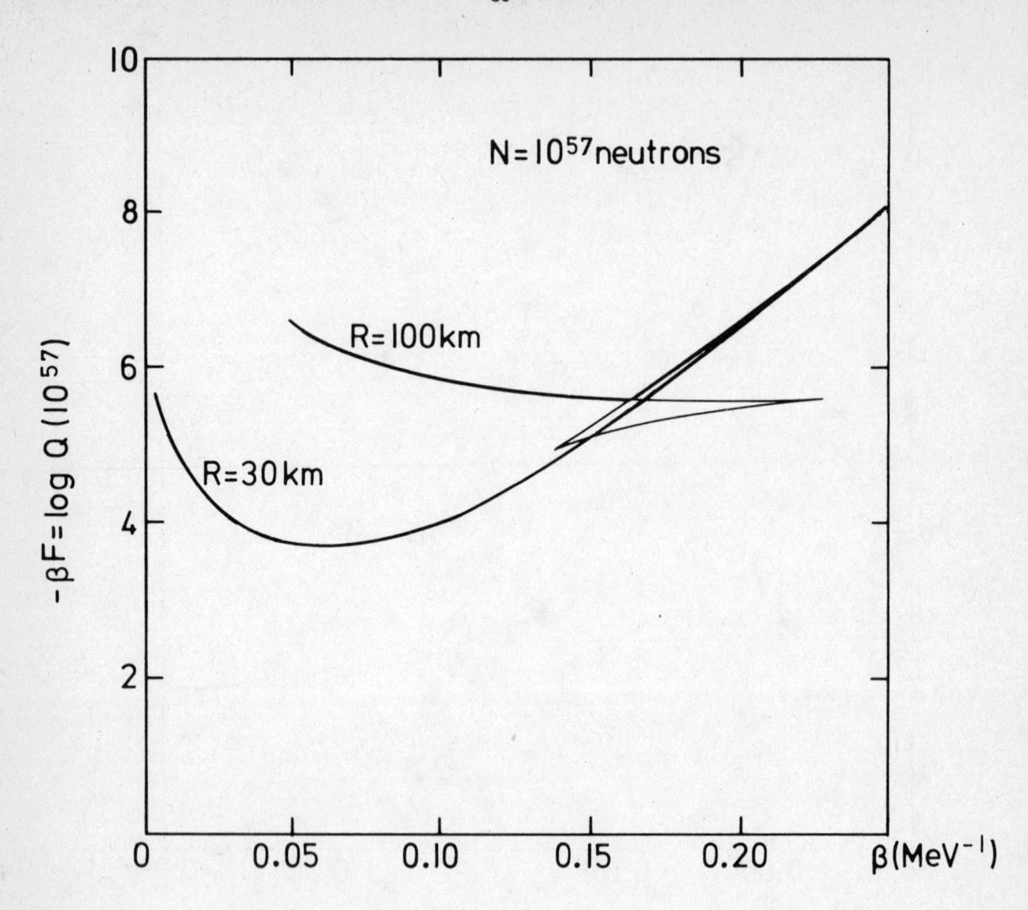

Ad 4.3): Free Energy/Temperature versus Inverse Temperature

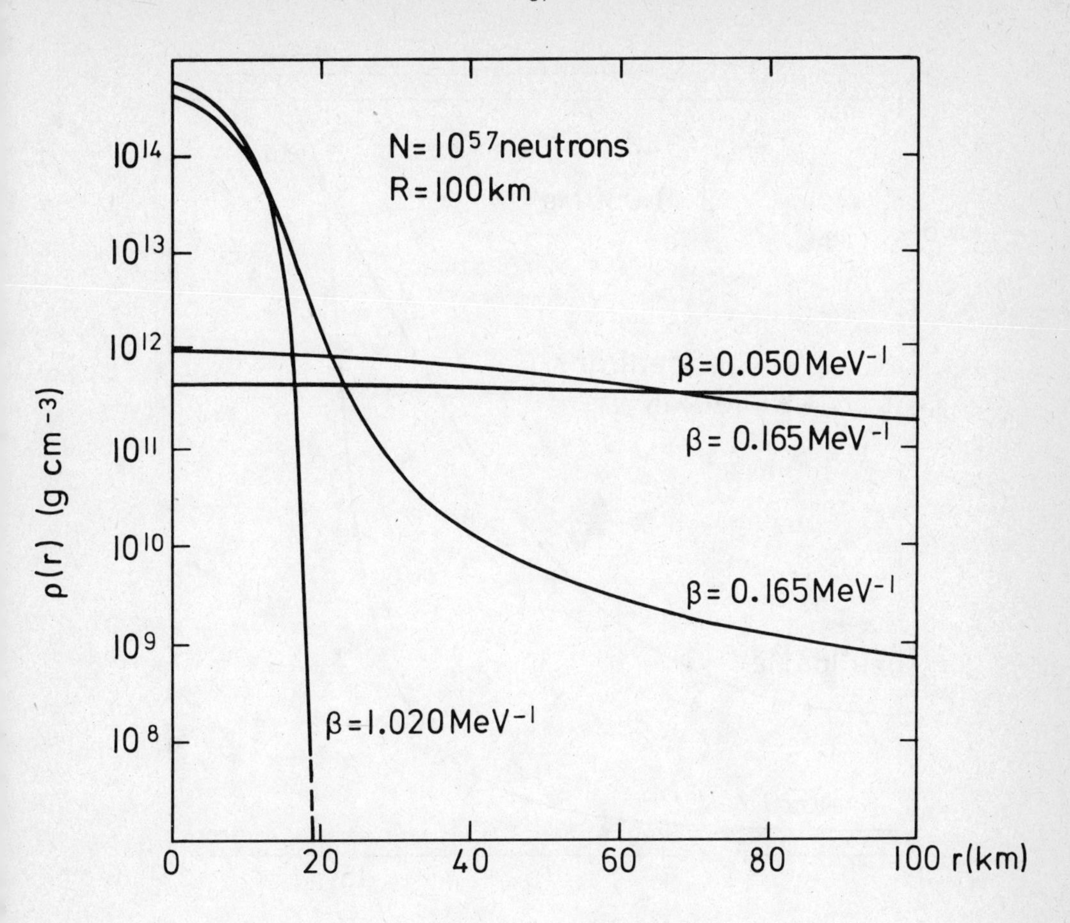

Ad 4.4): Mass Distribution versus Radial Distance

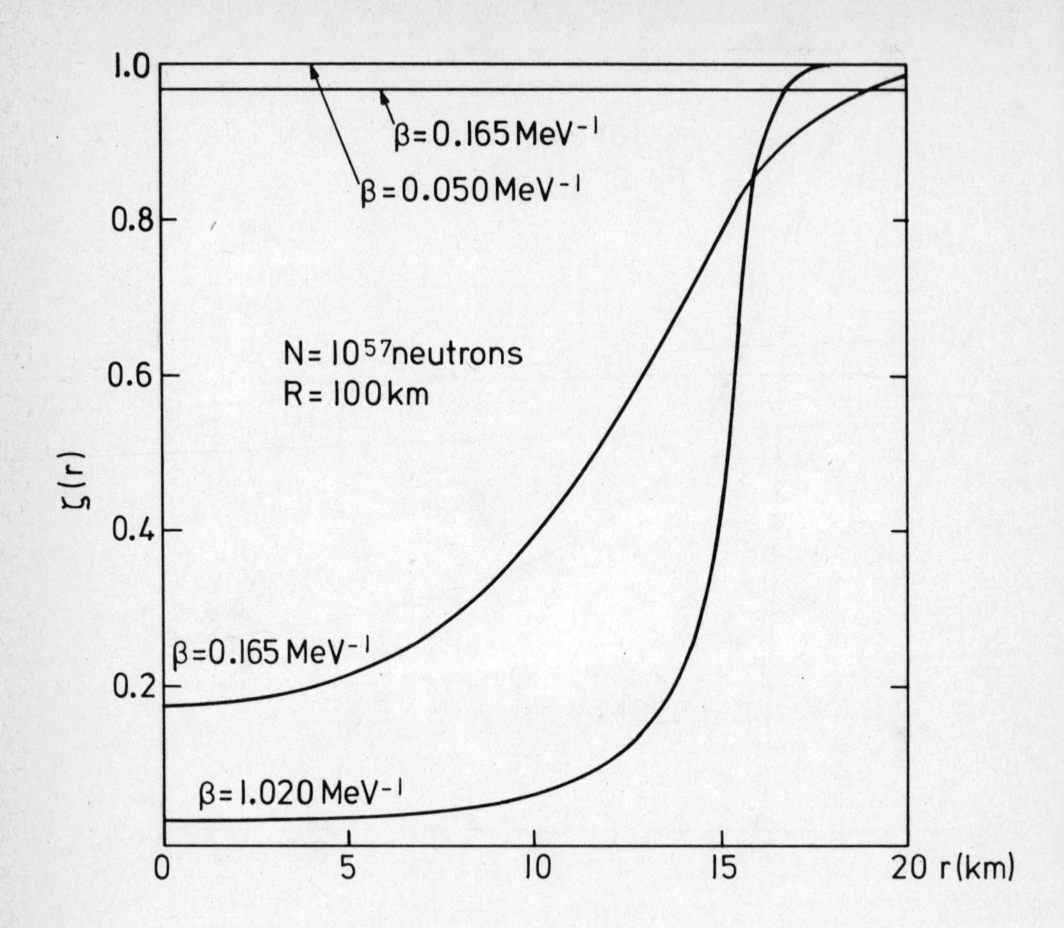

Ad 4.5): Fermi Degeneracy versus Radial Distance

Ad 4.6): Pressure versus Volume

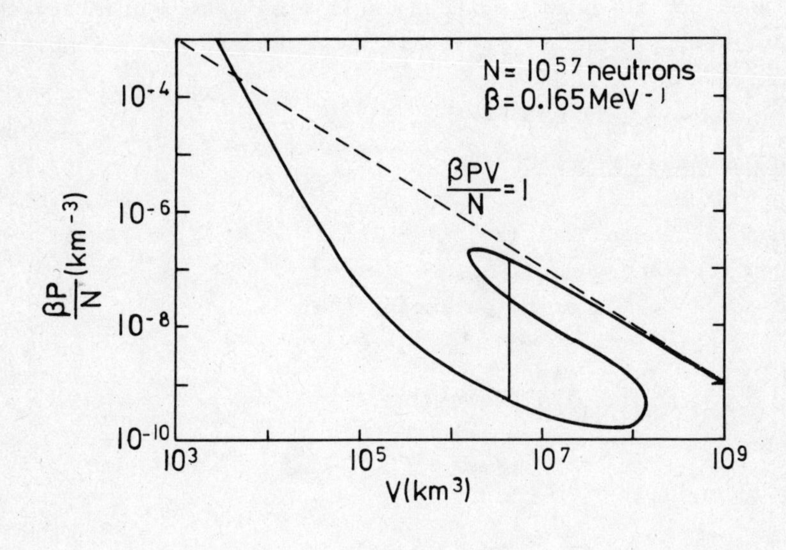

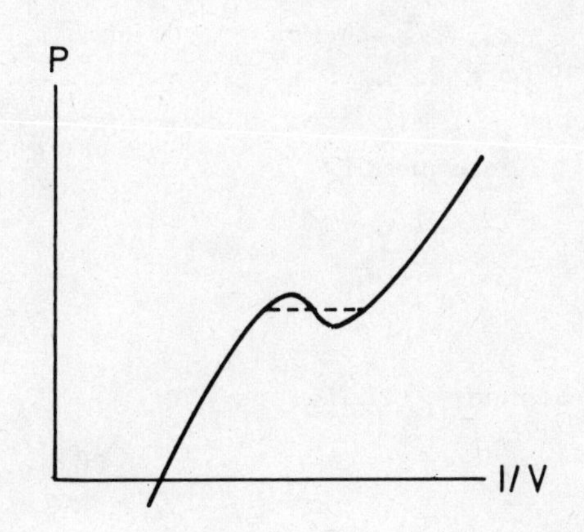

I.c. Grand Canonical Ensemble

The typical problem of the grand canonical ensemble is the following: We have used the particle number N as scaling parameter, and have let tend N to infinity. Grand canonically we have to sum over the N. Even this sum need not to exist, neither is it clear which quantity should tend to infinity. The main problem is to formulate a reasonable partition function.

I.ca. The Partition Function

Let $\ell > 0$ be fixed and $z = (z_1, z_2)$ (with $z_\alpha = e^{\beta\mu_\alpha}$) be the activities of the electrons and nuclei.
Let $\beta > 0$ and λ be a scaling parameter. Define

$$\beta p_\lambda(\beta,z) = \frac{1}{(\lambda^{1/3}\ell)^3} \ln \Xi(\lambda,\beta,z) \tag{391}$$

with

$$\Xi(\lambda,\beta,z) = \sum_{N_1=0}^{\infty} \sum_{N_2=0}^{\infty} z_1^{N_1} z_2^{N_2} \operatorname{Tr}_{\mathcal{H}_{N,\ell}} \exp(-\beta \tilde{H}_{N,\lambda}) . \tag{392}$$

The effective Hamiltonian $\tilde{H}_{N,\lambda}$ is determined according to the following requirements:

(a) If $\kappa = 0$ and $e^2 = 0$ then $\tilde{H}_{N,\lambda} = \hat{T}_{N,\lambda}$ and the pressure coincides in the thermodynamic limit with the ordinary pressure of the ideal Fermi gas. Consequently

$$\hat{T}_{N,\lambda} = \lambda^{-2/3} \sum_{\alpha i_\alpha} \frac{-\Delta_{\alpha i_\alpha}}{2M_\alpha} \tag{393}$$

with Dirichlet boundary conditions.

(b) Additivity:

$$\tilde{H}_{N,\lambda} = \hat{T}_{N,\lambda} + \frac{1}{2} f_1(N,\lambda) \sum_{\substack{\alpha\ i_\alpha \\ \beta\ k_\beta}}{}' \frac{(e_\alpha e_\beta - \kappa M_\alpha M_\beta)}{|x_{\alpha i_\alpha} - x_{\beta k_\beta}|} + f_2(N,\lambda) . \tag{394}$$

To fix the functions f_1 and f_2 we require:

(c) The pressure is dominated by $\lambda = N$ when $\lambda \to \infty$.
Comparison with (198-200) or (214-217) (for $\gamma = 0$) leads to

$$f_1(N,N) = \frac{1}{N}, \quad f_2(N,N) = 0. \tag{395}$$

(d) The simplest nontrivial effective Hamiltonian is consequently

$$\tilde{H}_{N,\lambda} = \lambda^{-2/3} \sum_{\alpha i_\alpha} \frac{-\Delta_{\alpha i_\alpha}}{2M_\alpha} + \frac{1}{2N} \sum_{\substack{\alpha\ i_\alpha \\ \beta\ k_\beta}}{}' \frac{(e_\alpha e_\beta - \kappa M_\alpha M_\beta)}{|x_{\alpha i_\alpha} - x_{\beta k_\beta}|} \tag{396}$$

with Dirichlet boundary conditions.

I.cb. The Thomas-Fermi Limit in the Grand Canonical Ensemble

Theorem (397): Let $\ell > 0$ be fixed and $\beta > 0$ and $z = (z_1, z_2)$ be given. Then the Thomas-Fermi limit

$$\Pi(\beta,z) = \lim_{\lambda\to\infty} p_\lambda(\beta,z) \tag{398}$$

exists, and is given explicitly by

$$\Pi(\beta,z) = \max_{\rho_\alpha} P_\rho[\rho_\alpha] \tag{399}$$

with

$$\ell^3 P_\rho = \int_{\Lambda_\ell} d^3x p_\rho(x) + u \tag{400}$$

if (β,z) is not a point where a phase transition (in the canonical ensemble) occurs. Here $p_\rho(x)$ is given by (380), u is given by (313,314), and ρ_α is determined by the self-consistency equation (315).
If $(\beta,z^{(1)},z^{(2)})$ is a point where a phase transition (in the canonical ensemble) occurs, then for each $\mu_\alpha = \beta^{-1}\log z_\alpha = a\mu_\alpha^{(1)} + (1-a)\mu_\alpha^{(2)}$, for some $0 \le a \le 1$, the limit pressure is

$$\Pi(\beta,z) = a\Pi(\beta,e^{\beta\mu^{(1)}}) + (1-a)\Pi(\beta,e^{\beta\mu^{(2)}}) . \tag{401}$$

Proof: For a one component system of purely cosmic matter parts of (397) have been proven in [41]. We conjecture that the generalization to ordinary cosmic matter, as stated here, remains true.

Remark: The temperature dependent Thomas-Fermi equation for the grand canonical ensemble - (315,316,317) and $\Pi(\beta,z) = P_\rho[\rho^o_\alpha]$ (in analogy to (319)) - becomes exactly valid in the Thomas-Fermi limit $\lambda\to\infty$.

II. THE THERMODYNAMIC THOMAS-FERMI LIMIT FOR EQUILIBRIUM STATES

II.a. Correlation Functions

We confine our treatment to the Thomas-Fermi limit for the correlation functions in the canonical ensemble with γ chosen to be zero.

II.aa. The Generalized Free Energy

II.aa.1. The Basic Concept

The n-point correlation functionals applied to test functions $f_i \in \mathcal{S}(\mathbb{R}^3)$ are the mean values of the operators of the integrated n-point correlation functionals

(402) $$\hat{\rho}_N^{\underline{\alpha}}(f_1,\ldots,f_n) = \Big(\prod_{k=1}^{n} N_{\alpha_k}^{-1} \sum_{i_{\alpha_k}=1}^{N_{\alpha_k}}\Big) S(x_{i_{\alpha_1}},\ldots,x_{i_{\alpha_n}}) \ ,$$

where

(403) $$S(x_1,\ldots,x_n) = n!^{-1} \sum_{\pi\in S_n} \prod_{\delta=1}^{n} f_\delta(x_{\pi(\delta)})$$

is the symmetrized product of n test functions and $\underline{\alpha} = (\alpha_1,\ldots,\alpha_n)$. The operators of the integrated n-point correlation functionals are bounded and the equilibrium expectation values are

(404) $$\langle\hat{\rho}_N^{\underline{\alpha}}(f_1,\ldots,f_n)\rangle_\beta = \mathrm{Tr}_{\mathcal{H}_{N,\ell}} \{\hat{\rho}_N^{\underline{\alpha}}(f_1,\ldots,f_n)\, e^{-\beta\tilde{H}_N}\} \exp(\beta N\Phi(N,\beta,\ell))$$

$$= \frac{\partial}{\partial\lambda}\, \Phi^\lambda(N,\beta,\ell,\lambda)\big|_{\lambda=0}$$

with the local generalized free energy

(405) $$\Phi^\lambda(N,\beta,\ell,\lambda) = -\frac{1}{\beta N}\,\ell n\, \mathrm{Tr}_{\mathcal{H}_{N,\ell}} \exp(-\beta\{\tilde{H}_N + \lambda N\hat{\rho}_N^{\underline{\alpha}}(f_1,\ldots,f_n)\})$$

and

(406) $\Phi^{\lambda}(N,\beta,\ell,0) = \Phi(N,\beta,\ell)$.

Proving the existence of the thermodynamic limit $N \to \infty$ of the correlation functions (404) consists in showing the existence of the limit of the local generalized free energy and the equality of the limit of the derivative with the derivative of the limiting generalized free energy.

II.aa.2. The Discretization

We generalize the discretization procedure outlined in I.bb. $S_{\mu g}(x_1,\ldots,x_n)$ are defined to be step functions, which are constant in each n-tuple of cells and approximate $S(x_1,\ldots,x_n)$. S_μ can be taken equal to S. Let the index σ stand for $\mu g, \mu$ or none.

(407) $$\hat{H}_{N,\sigma}(\lambda) := \hat{T} + \hat{V}_\sigma + \lambda N\Big(\prod_{k=1}^{n} N_{\alpha_k}^{-1} \sum_{i_{\alpha_k}=1}^{N_{\alpha_k}}\Big) S_\sigma(x_{i_{\alpha_1}},\ldots,x_{i_{\alpha_n}}) ,$$

(408) $$\Phi_\sigma^\lambda(N,\beta,\ell,\lambda) := -\frac{1}{\beta N} \ln \mathrm{Tr}_{\mathcal{H}_{N,\ell}} \exp(-\beta \hat{H}_{N,\sigma}(\lambda)) .$$

Lemma (409): The following Thomas-Fermi limit exists

(410) $$\lim_{N\to\infty} \Phi_{\mu g}^\lambda(N,\beta,\ell,\lambda) = \Phi_{\mu g}^\lambda(\beta,\ell,\lambda)$$

with

(411) $$\Phi_{\mu g}^\lambda = \frac{1}{2}\sum_\alpha \int_{\Lambda_\ell} d^3x \rho_{\alpha,\mu g}(x) W_{\alpha,\mu g}(x) +$$
$$+ \lambda \int_{\Lambda_\ell^n} S_{\mu g}(x_1,\ldots,x_n) \prod_{k=1}^{n} \rho_{\alpha_k,\mu g}(x_k) d^3x_k +$$
$$+ \sum_\alpha \int_{\Lambda_\ell} d^3x\, \rho_{\alpha,\mu g}(x) \mu_{\alpha,\mu g}(x) -$$
$$- \beta^{-1} \sum_\alpha \int_{\Lambda_\ell} d^3x \int q \frac{d^3p}{(2\pi)^3} \ln \{1 + e^{-\beta(\frac{p^2}{2M_\alpha} - \mu_{\alpha,\mu g}(x))}\} .$$

The ρ's and μ's are solutions of the generalized self-consistency equations

$$(412) \qquad \mu_{\alpha,\mu g}(x) = \mu_\alpha - W_{\alpha,\mu g}(x) - \lambda S_{\alpha,\mu g}(x) \ ,$$

$$(413) \qquad \rho_{\alpha,\mu g}(x) = q \int \frac{d^3p}{(2\pi)^3} \left(1 + e^{\beta(\frac{p^2}{2M_\alpha} - \mu_{\alpha,\mu g}(x))}\right)^{-1} \ ,$$

$$(414) \qquad \int_{\Lambda_\ell} \rho_{\alpha,\mu g}(x) d^3x = \zeta_\alpha \ ,$$

with

$$(415) \qquad S_{\alpha,\mu g}(x) = n \int_{\Lambda_\ell^{n-1}} S_{\mu g}(x,x_2,\ldots,x_n) \prod_{k=2}^{n} \rho_{\alpha_k,\mu g}(x_k) d^3x_k \ ,$$

and

$$(416) \qquad \zeta_1 = 1, \quad \zeta_2 = \zeta \ .$$

If there are more than one solution of the generalized self-consistency equations the ones which minimize $\Phi^\lambda_{\mu g}$ must be chosen.

Proof: For purely cosmic matter it has been observed in [42] that the proof follows without complications from the case $\lambda=0$ treated in [34]. Our generalization to ordinary cosmic matter follows by inspection from the procedure for $\lambda=0$ presented in Chapter I.bb. □

Lemma (417): The following thermodynamic Thomas-Fermi limit exists

$$(418) \qquad \lim_{N\to\infty} \Phi^\lambda(N,\beta,\ell,\lambda) = \lim_{\mu\to\infty} \lim_{g\to\infty} \Phi^\lambda_{\mu g}(\beta,\ell,\lambda) \ .$$

Proof: This fact has been observed first in [42] for purely cosmic matter, and it is, in the generalization stated here, again a consequence of Chapter I.bb.

II.aa.3. Continuity of the Generalized Thomas-Fermi Equation

Definitions (419):

(420) $$\Phi^{\lambda}(\beta,\ell,\lambda) = \min_{\rho_\alpha \in \Omega_\alpha} F^{\lambda}_{\rho}[\rho_\alpha](\beta,\ell,\lambda) ,$$

(421) $$F^{\lambda}_{\rho,\sigma}[\rho_\alpha] = \frac{1}{2}\sum_{\alpha}\int_{\Lambda_\ell} \rho_\alpha(x) W_{\alpha,\sigma}[\rho_\alpha](x) d^3x + $$
$$+ \lambda\int_{\Lambda_\ell^n} S_\sigma(x_1,\dots,x_n) \prod_{k=1}^{n} \rho_{\alpha_k}(x_k) d^3x_k + \sum_{\alpha}\int_{\Lambda_\ell} \rho_\alpha(x)\mu_\alpha(x) d^3x$$
$$- \beta^{-1}\sum_{\alpha}\int_{\Lambda_\ell} d^3x \int q \frac{d^3p}{(2\pi)^3} \ln\{1 + e^{-\beta(\frac{p^2}{2M_\alpha} - \mu_\alpha(x))}\} ,$$

(compare with (312)),

(422) $$\Omega_\alpha = \{\rho_\alpha \in \mathcal{L}_1(\Lambda_\ell)/\rho_\alpha(x) \geq 0, \int_{\Lambda_\ell} \rho_\alpha(x) d^3x = \zeta_\alpha\} .$$

The $\mu_\alpha(x)$ is here a functional of $\rho_\alpha(x)$, implicitly defined by (413) with the index μg dropped.

Lemma (423): If $\rho_{\alpha,\sigma}(x)$ is a solution of the generalized self-consistency equations for β, ζ_α and ℓ, then there exist numbers $\hat{\mu}_\alpha$, $d_{\alpha,q}$, and k_α such that

(424) $$\mu_\alpha \leq \hat{\mu}_\alpha ,$$

(425) $$\|\rho_{\alpha,\sigma}\|_q \leq d_{\alpha,q} \qquad \text{for } q < 2 ,$$

and

(426) $$|\rho_{\alpha,\sigma}(x)| \leq (2M_\alpha)^{3/2} g_\alpha(\beta,\hat{\mu}_\alpha + \bar{c}_p\bar{d}_q + |\lambda| k_\alpha)$$

with

$$(427) \qquad g_\alpha(\beta, y_\alpha) = \int q \frac{d^3p}{(2\pi)^3} \left(1 + e^{\beta(p^2 - y_\alpha)}\right)^{-1} ,$$

and

$$(428) \qquad \bar{c}_p = (Z^2e^2 + \kappa M^2) \, \| \tfrac{1}{|x|} \|_p , \quad \bar{d}_q = \sum_\alpha d_{\alpha,q}$$

for $2 < p < 3$; $\frac{3}{2} < q < 2$; $p^{-1} + q^{-1} = 1$.

<u>Proof</u>: Except for minor alterations the proof is a generalization of the arguments given in [42]. If $p < 3$ and $\tilde{c}_p = \|\frac{1}{|x|}\|_p$ is the $\mathcal{L}_p(\Lambda_\ell)$-norm, then the function $v_{\alpha\beta} : z \to v_{\alpha\beta}(z) = |z|^{-1}(e_\alpha e_\beta - \kappa M_\alpha M_\beta)$ obeys

$$(429) \qquad \| v_{\alpha\beta}(x - \cdot) \|_p \leq \| v_{\alpha\beta} \|_p \leq (Z^2e^2 + \kappa M^2)\tilde{c}_p .$$

Now we study bounds to the solutions of the generalized self-consistency equations (412,413,414) with the index μg substituted by σ:

$$(430) \qquad \rho_{\alpha,\sigma}(x) = (2M_\alpha)^{3/2} g_\alpha(\beta, \mu_{\alpha,\sigma}(x)) ,$$

$$(431) \qquad \mu_{\alpha,\sigma}(x) = \mu_\alpha - W_{\alpha,\sigma}(x) - \lambda S_{\alpha,\sigma}(x) ,$$

$$(432) \qquad \int_{\Lambda_\ell} \rho_{\alpha,\sigma}(x) d^3x = \zeta_\alpha .$$

It is straightforward to verify:

$$(433) \qquad g_\alpha(\beta, \mu_\alpha) \leq \Theta(1 - \beta\mu_\alpha)\beta^{-3/2} g_\alpha(1,1) + \Theta(\beta\mu_\alpha - 1)(6\pi^2)^{-1}(\mu_\alpha^{3/2} + \frac{3}{2}\mu_\alpha\beta^{-1/2} + \frac{3}{2}\beta^{-3/2}) .$$

Therefore $\rho_{\alpha,\sigma}$ as functional of $\mu_{\alpha,\sigma}$ does not grow stronger than $\mu_{\alpha,\sigma}^{3/2}$.

Since $W_{\alpha,\sigma}$ is proportional to a convex combination of the $v_{\alpha\beta,\sigma}$, and (429) holds also for $v_{\alpha\beta}$ replaced by $v_{\alpha\beta,\sigma}$, we have

$$(434) \qquad \|W_{\alpha,\sigma}\|_p \leq \zeta_\alpha(Z^2e^2 + \kappa M^2)\tilde{c}_p = \bar{c}_p\zeta_\alpha$$

for $p < 3$. Similarly, but for all p, including $p=\infty$, there is a number k_α, such that

$$(435) \qquad \|S_{\alpha,\sigma}\|_p \leq k_\alpha \ .$$

By (433),(430) and by (431),(434),(435) the $\rho_{\alpha,\sigma}$ are in a bounded set in every $\mathcal{L}_{2p/3}(\Lambda_\ell)$. Consequently there exists a number $d_{\alpha,q}$ for every $q < 2$ with

$$(436) \qquad \|\rho_{\alpha,\sigma}\|_q \leq d_{\alpha,q} \ .$$

To prove (424) we consider $\bar{\mu}_{\alpha,\sigma}(x) = \mu_{\alpha,\sigma}(x) - \mu_{\alpha,\sigma}$. Without restriction of generality we can assume that the set of admitted λ's is bounded from above by some number λ_1. Let

$$(437) \qquad m_\alpha = \bar{c}_p\zeta_\alpha + \lambda_1 k_\alpha \ ,$$

then by (434) and (435) the uniform bound

$$(438) \qquad \|\bar{\mu}_{\alpha,\sigma}\|_p \leq m_\alpha$$

holds. Define $b_\alpha \in \mathbb{R}$ and the set $V_{\alpha,\sigma} \subset \Lambda_\ell$ by

$$(439) \qquad \frac{1}{2}\,\ell^3 b_\alpha^p = m_\alpha^p \ ,$$

$$(440) \qquad V_{\alpha,\sigma} = \{x \in \Lambda_\ell / |\bar{\mu}_{\alpha,\sigma}(x)| \leq b_\alpha\} \ .$$

Since

$$(441) \qquad \frac{1}{2}\,\ell^3 b_\alpha^p \geq \|\bar{\mu}_{\alpha,\sigma}\|_p^p \geq \int_{\Lambda_\ell \setminus V_{\alpha,\sigma}} |\bar{\mu}_{\alpha,\sigma}(x)|^p d^3x > b_\alpha^p(\ell^3 - |V_{\alpha,\sigma}|) \ ,$$

we have for the volume of $V_{\alpha,\sigma}$:

$$(442) \qquad |V_{\alpha,\sigma}| > \frac{1}{2}\,\ell^3 \ .$$

Now let

$$(443) \qquad A_{\alpha,\sigma} = \{x \in \Lambda_\ell / \rho_{\alpha,\sigma}(x) > (2M_\alpha)^{3/2} g_\alpha(\beta,\mu_\alpha - b_\alpha)\} .$$

If $x \in V_{\alpha,\sigma}$, then

$$(444) \qquad \rho_{\alpha,\sigma}(x) = (2M_\alpha)^{3/2} g_\alpha(\beta,\mu_\alpha + \bar{\mu}_{\alpha,\sigma}(x)) > (2M_\alpha)^{3/2} g_\alpha(\beta,\mu_\alpha - b_\alpha),$$

and therefore

$$(445) \qquad |A_{\alpha,\sigma}| \geq |V_{\alpha,\sigma}| > \frac{1}{2}\,\ell^3 \ .$$

Let a $\hat{\mu}_\alpha$ be defined by

$$(446) \qquad \frac{1}{2}\,\ell^3 (2M_\alpha)^{3/2} g_\alpha(\beta,\hat{\mu}_\alpha - b_\alpha) = \zeta_\alpha \ .$$

$\hat{\mu}_\alpha$ is uniquely defined by monotonicity and continuity of $g_\alpha(\beta,\cdot)$.

Since

$$(447) \qquad \zeta_\alpha = \int_{\Lambda_\ell} \rho_{\alpha,\sigma}(x) d^3x > \int_{A_{\alpha,\sigma}} \rho_{\alpha,\sigma}(x) d^3x > \frac{1}{2}\,\ell^3 (2M_\alpha)^{3/2} g_\alpha(\beta,\mu_\alpha - b_\alpha),$$

it follows necessarily from the monotonicity of $g_\alpha(\beta,\cdot)$ that all possible μ_α are bounded from above by $\hat{\mu}_\alpha$.

Finally we apply Hölder's inequality to $|W_{\alpha,\sigma}(x)|$, to obtain

$$(448)\qquad \sup_{x\in\Lambda_\ell} \Big|\sum_\beta \int v_{\alpha\beta,\sigma}(x-y)\rho_{\beta,\sigma}(y)d^3y\Big| \leq$$

$$\leq \sup_{x\in\Lambda_\ell} \|v_{\alpha\beta,\sigma}(x-\cdot)\|_p \sum_\alpha \|\rho_{\alpha,\sigma}\|_q \leq \bar{c}_p\bar{d}_q$$

by (429), (and $\|v_{\alpha\beta,\sigma}\|_p \leq \bar{c}_p$ in general), (436), (428), and with the p's and q's are chosen as indicated in (423). The bounds (448), (424), and (435) together lead in (430) to the result (426) if monotonicity of $g_\alpha(\beta,\cdot)$ is used. Notice that $d_{\alpha,q}$ is defined consistently by first applying (424) to (433), and then using (438). □

<u>Lemma</u> (449): Let $\rho_{\alpha,\mu g}$ be a solution of the generalized self-consistency equations for the discretized formulation (412-414), then there exists a strongly on $(\Omega_\alpha, \|\cdot\|_\infty)$ converging subsequence when first g and then μ tend to infinity.

<u>Proof</u>: The proof follows the ideas sketched in [42]. Define

$$(450)\qquad \bar{W}_{\alpha,\mu g}(x) = \sum_\beta \int_{\Lambda_\ell} d^3y\; v_{\alpha\beta,\mu}(x-y)\rho_{\beta,\mu g}(y) \quad ,$$

$$(451)\qquad T_{\alpha,\mu g}(x) = n \int_{\Lambda_\ell^{n-1}} S(x,x_2,\dots,x_n) \prod_{k=2}^{n} \rho_{\alpha_k,\mu g}(x_k)d^3x_k \quad .$$

Consider the subset of the real Banach space (with the $\mathcal{L}_\infty$-topology) of continuous functions on the compact metric space $\bar{\Lambda}_\ell$, which is formed by all $\bar{W}$ and T. This subset is equicontinuous, which follows immediately for the T's, and for the $\bar{W}$'s because

$$(452)\qquad |\nabla\bar{W}_{\alpha,\mu g}(x)| = \Big|\sum_\beta \int_{\Lambda_\ell} d^3y\; \rho_{\beta,\mu g}(y)\nabla_x v_{\alpha\beta,\mu}(x-y)\Big|$$

$$\leq \sum_\beta \|\rho_{\beta,\mu g}\|_\infty (Z^2e^2 + \kappa M^2) \int_{\Lambda_\ell} \Big|\nabla \frac{1}{|x|}\Big| d^3x \quad ,$$

and the right-hand side is bounded independently of α, μ, g. By the theorem of Ascoli the set of functions $\bar{W}$, T is relatively compact, and the sequences of functions $\overline{W}_{\alpha,\mu g}$ and $T_{\alpha,\mu g}$ therefore have subsequences, which are strongly convergent with respect to the $\mathcal{L}_\infty$-topology. If first $g \to \infty$, and then $\mu \to \infty$, then there exist also strongly convergent subsequences of $W_{\alpha,\mu g}$, which can be proven twice. First it follows, because the $v_{\alpha\beta,\mu g}$ converge in $\mathcal{L}_1(\overline{\Lambda}_\ell - \overline{\Lambda}_\ell)$-norm to $v_{\alpha\beta,\mu}$ for $g \to \infty$ by Lebesgue's convergence theorem. Secondly it follows, because by the Stone-Weierstrass approximation theorem $v_{\alpha\beta,\mu}$ is uniformly approximated by $v_{\alpha\beta,\mu g}$. The second reasoning can be used again to show the existence of strongly converging subsequences for $S_{\mu g}$. The chemical potentials μ_α, although not indicated explicitly in (412), depend also on μ and g via (414). They are bounded from above (independently of μ and g) by (424), and they are bounded from below (independently of μ and g) because of the inequality

$$\zeta_\alpha \leq (2M_\alpha)^{3/2}\beta^{-3/2}\ell^3\{q \int \frac{d^3\overline{p}}{(2\pi)^3} e^{-\overline{p}^2}\} e^{\beta(\overline{c}_p\overline{d}_q+\lambda_1 k_\alpha)} e^{\beta\mu_\alpha} , \tag{453}$$

where the notations of Lemma (423) are used. Now we repeat the argumentation with the subset of continuous functions $\mu_\alpha - \overline{W}_{\alpha,\mu g} - \lambda T_{\alpha,\mu g}$ and by Ascoli's theorem (and by the Borel-Lebesgue theorem) there exists a (common) strongly convergent subsequence of the functions $\mu_{\alpha,\mu g} = \mu_\alpha - W_{\alpha,\mu g} - \lambda S_{\alpha,\mu g}$, if first $g \to \infty$, and then $\mu \to \infty$. Let the function $\tilde{\mu}_\alpha$ denote the strong limit point. From (430) or (413) we infer

$$\|\rho_{\alpha,\mu g} - \tilde{\rho}_\alpha\|_\infty \leq \max(\|\rho_{\alpha,\mu g}\|_\infty, \|\tilde{\rho}_\alpha\|_\infty)\ \beta\|\mu_{\alpha,\mu g} - \tilde{\mu}_\alpha\|_\infty \tag{454}$$

with

$$\tilde{\rho}_\alpha(x) = (2M_\alpha)^{3/2} g_\alpha(\beta, \tilde{\mu}_\alpha(x)). \tag{455}$$

Since the bound (426) holds also for $\tilde{\rho}_\alpha(x)$, there exists a subsequence, such that $\rho_{\alpha,\mu g}$ converges strongly to $\tilde{\rho}_\alpha$ if first $g \to \infty$, and then $\mu \to \infty$. □

Proposition (456): $\Phi^\lambda(\beta,\ell,\lambda) = \lim_{\mu\to\infty} \lim_{g\to\infty} \Phi^\lambda_{\mu g}(\beta,\ell,\lambda)$.

Proof: We follow [42]. Let $\rho_\alpha \in \Omega_\alpha$ be fixed. It follows immediately from (421) that the following limit exists:

$$(457) \qquad F_\rho^\lambda[\rho_\alpha] = \lim_{\mu\to\infty} \lim_{g\to\infty} F_{\rho,\mu g}^\lambda[\rho_\alpha] .$$

Therefore one has

$$(458) \qquad \min_{\rho_\alpha \in \Omega_\alpha} F_\rho^\lambda[\rho_\alpha](\beta,\ell,\lambda) \geq \lim_{\mu\to\infty} \lim_{g\to\infty} \min_{\rho_\alpha \in \Omega_\alpha} F_{\rho,\mu g}^\lambda[\rho_\alpha](\beta,\ell,\lambda).$$

The densities $\rho_{\alpha,\mu g}$ which minimize the functional $F_{\rho,\mu g}^\lambda[\cdot](\beta,\ell,\lambda)$ are solutions of the generalized self-consistency equations (412-414). With Lemma (449) there exist subsequences $\rho_{\alpha,\mu_i g_j}$, such that the strong limit

$$(459) \qquad \text{s-}\lim_{\mu_i\to\infty} \lim_{g_j\to\infty} \rho_{\alpha,\mu_i g_j} = \tilde{\rho}_\alpha$$

exists. With Lemma (449) it is easy to see from (421) that

$$(460) \qquad \lim_{\mu_i\to\infty} \lim_{g_j\to\infty} F_{\rho,\mu_i g_j}^\lambda[\rho_{\alpha,\mu_i g_j}](\beta,\ell,\lambda) = F_\rho^\lambda[\tilde{\rho}_\alpha](\beta,\ell,\lambda).$$

Consequently from (458)

$$(461) \qquad \Phi^\lambda(\beta,\ell,\lambda) = \min_{\rho_\alpha \in \Omega_\alpha} F_\rho^\lambda[\rho_\alpha](\beta,\ell,\lambda) =$$

$$= F_\rho^\lambda[\tilde{\rho}_\alpha](\beta,\ell,\lambda) = \lim_{\mu_i\to\infty} \lim_{g_j\to\infty} \Phi_{\mu_i g_j}^\lambda(\beta,\ell,\lambda),$$

because according to Lemma (409)

$$(462) \qquad \Phi_{\mu g}^\lambda(\beta,\ell,\lambda) = \min_{\rho_\alpha \in \Omega_\alpha} F_{\rho,\mu g}^\lambda[\rho_\alpha](\beta,\ell,\lambda) .$$

By Lemma (417) the limit

$$(463) \qquad \lim_{\mu\to\infty} \lim_{g\to\infty} \Phi_{\mu g}^\lambda(\beta,\ell,\lambda) = \lim_{\mu_i\to\infty} \lim_{g_j\to\infty} \Phi_{\mu_i g_j}^\lambda(\beta,\ell,\lambda)$$

exists. This concludes the proof. □

Remark (464): The last proofs demonstrates the "continuity of the generalized Thomas-Fermi equation". They show that the limits of the solutions of the generalized Thomas-Fermi equation for the discretized formulation are again solutions of the generalized Thomas-Fermi equation for the continuum formulation.

II. ab. Differentiability of the Generalized Free Energy

II.ab.1. Two Properties of Concave Functions

Lemma (465): If

a) $f(\lambda) = \min_{\rho_\alpha \in \Omega_\alpha} F[\underline{\rho}](\lambda)$, $(\underline{\rho} = (\rho_1, \ldots, \rho_n)$, $\alpha \in \{1,2,\ldots,n\})$

is a concave function such that the minimum is attained at some $\underline{\rho}_\lambda$:

$f(\lambda) = F[\underline{\rho}_\lambda](\lambda)$ and if

b) the partial derivatives

$$F'[\underline{\rho}](\lambda_o) = \frac{\partial}{\partial\lambda} F[\underline{\rho}](\lambda)\Big|_{\lambda=\lambda_o}$$

exist for all pairs $(\underline{\rho}, \lambda_o)$,

then the following relation holds

$$f'_+(\lambda) \leq F'[\underline{\rho}_\lambda](\lambda) \leq f'_-(\lambda),$$

(where $f'_+(f'_-)$ denotes the right- (left-) hand side derivative).

Proof: This lemma is a generalization of [42]. Since f is concave it is a standard result [43] that the derivatives f'_+ and f'_- exist. Now the lemma follows from the following estimation (for $\varepsilon > 0$):

$$(466) \qquad \begin{aligned} f'_+(\lambda) &= \lim_{\varepsilon\to 0} \frac{1}{\varepsilon} (F[\underline{\rho}_{\lambda+\varepsilon}](\lambda+\varepsilon) - F[\underline{\rho}_\lambda](\lambda)) \\ &\leq \lim_{\varepsilon\to 0} \frac{1}{\varepsilon} (F[\underline{\rho}_\lambda](\lambda+\varepsilon) - F[\underline{\rho}_\lambda](\lambda)) \\ &= F'[\underline{\rho}_\lambda](\lambda) = \\ &= \lim_{\varepsilon\to 0} \frac{1}{\varepsilon} (F[\underline{\rho}_\lambda](\lambda) - F[\underline{\rho}_\lambda](\lambda-\varepsilon)) \\ &\leq \lim_{\varepsilon\to 0} \frac{1}{\varepsilon} (F[\underline{\rho}_\lambda](\lambda) - F[\underline{\rho}_{\lambda-\varepsilon}](\lambda-\varepsilon)) \\ &= f'_-(\lambda). \end{aligned}$$

□

Lemma (467): If there exists a topology τ on $\Omega_1 \times \ldots \times \Omega_n$ such that
a) $F'[\underline{\rho}](\lambda)$ is continuous on $(\Omega_1 \times \ldots \times \Omega_n \times \mathbb{R}, \tau \times |\cdot|)$ and if
b) there exists a path $\lambda \in (\mathbb{R},|\cdot|) \to \underline{\rho}_\lambda \in (\Omega_1 \times \ldots \times \Omega_n, \tau)$, minimizing $F[\cdot](\lambda)$, which is continuous at λ_o, then $f(\lambda)$ is differentiable at λ_o and

$$f'(\lambda_o) = F'[\underline{\rho}_{\lambda_o}](\lambda_o) \; .$$

Proof: The proof is a generalization of [42]. Since f is concave, the following relations hold: For every $x > y$ we have $f'_+(x) \leq f'_-(x) \leq f'_+(y)$. With Lemma (465): Let $\varepsilon > 0$, then

$$(468) \qquad F'[\underline{\rho}_{\lambda_o+\varepsilon}](\lambda_o+\varepsilon) \leq f'_-(\lambda_o+\varepsilon) \leq f'_+(\lambda_o) \leq F'[\underline{\rho}_{\lambda_o}](\lambda_o) \leq f'_-(\lambda_o)$$

$$\leq f'_+(\lambda_o-\varepsilon) \leq F'[\underline{\rho}_{\lambda_o-\varepsilon}](\lambda_o-\varepsilon) .$$

Now apply $\lim_{\varepsilon \to 0}$ to (468), then we conclude with τ-continuity:

$$(469) \qquad \lim_{\varepsilon \to 0} F'[\underline{\rho}_{\lambda_o \pm \varepsilon}](\lambda_o \pm \varepsilon) = F'[\tau\text{-}\lim_{\varepsilon \to 0}(\underline{\rho}_{\lambda_o \pm \varepsilon}, \lambda_o \pm \varepsilon)] =$$

$$= F'[\underline{\rho}_{\lambda_o}](\lambda_o) = f'_+(\lambda_o) = f'_-(\lambda_o) \; ,$$

and the left-hand side coincide with the right-hand side in (468), after the limit $\varepsilon \to 0$ is carried out. □

II.ab.2. The Cluster Property

We apply (465) and (467) (for n=2) to compute the limit of the derivative of the generalized free energy with respect to λ at $\lambda_o = 0$.

Lemma (470): Let $\tilde{\underline{\rho}}(\lambda)$ be a solution of the generalized Thomas-Fermi equation ((412-414) with the index μg dropped and $\tilde{\underline{\rho}}(\lambda)$ is minimizing $F^\lambda_\rho[\underline{\rho}]$). If $\underline{\rho}^{TF}$ is a unique solution of the Thomas-Fermi equation ((315-317) and $\underline{\rho}^{TF}$ is minimizing $F_\rho[\underline{\rho}]$) then

$$\text{s-lim}_{\lambda\to 0} \underline{\tilde{\rho}}(\lambda) = \underline{\rho}^{TF}$$

exists, and $\lambda \in (\mathbb{R},|\cdot|) \to \underline{\tilde{\rho}}(\lambda) \in (\Omega_1 \times \Omega_2, \|\cdot\|_\infty \times \|\cdot\|_\infty)$ is continuous at $\lambda=0$.

<u>Proof</u>: The chemical potential μ_α is a functional of $\tilde{\rho}(\lambda)$ by (412-414) (with the index μg dropped). Using the boundedness of the μ_α, Lebesgue convergence theorem, and the continuity and monotonicity of $g_\alpha(\beta,\cdot)$ it is immediate that the function $\lambda \to \mu_\alpha[\tilde{\rho}_\alpha(\lambda)]$ is continuous. It is easy to see that the functional $W_\alpha[\rho_\alpha]$ is strongly continuous on $(\Omega_\alpha,\|\cdot\|_\infty)$. Therefore, let λ_i be any arbitrary subsequence converging to zero, and let $\underline{\tilde{\rho}}(\lambda_i)$ converge strongly to $\underline{\tilde{\rho}}'$, then $\mu_\alpha[\tilde{\rho}_\alpha(\lambda_i)](x)$ converges strongly to $\mu_\alpha[\tilde{\rho}_\alpha'](x) = \mu_\alpha[\tilde{\rho}_\alpha'] - W_\alpha[\tilde{\rho}_\alpha'](x)$, where the limit $\mu_\alpha[\tilde{\rho}_\alpha'] = \lim_{\lambda_i\to 0} \mu_\alpha[\tilde{\rho}_\alpha(\lambda_i)]$ exists. By Lebesgue's convergence theorem the strong limit point $\underline{\tilde{\rho}}'$ is a solution of the self-consistency equation (315-317), as can be seen by performing the limit $\lambda_i \to 0$ in the equations (412-414) (with the index μg dropped). Since $F_\rho^\lambda[\underline{\rho}_\lambda](\beta,\ell,\lambda)$ has by inspection the desired $\|\cdot\|_\infty$-continuity property

$$(471) \qquad \lim_{\lambda\to 0} F_\rho^\lambda[\underline{\rho}_\lambda](\beta,\ell,\lambda) = F_\rho^\lambda[\text{s-lim}_{\lambda\to 0}\, \underline{\rho}_\lambda](\beta,\ell,0),$$

and $f(\lambda) = \min_{\rho_\alpha\in\Omega_\alpha} F_\alpha^\lambda[\underline{\rho}](\beta,\ell,\lambda)$ as a concave function in some interval $[-\lambda_1,\lambda_1]$ is continuous at $\lambda=0$, the strong limit point $\underline{\tilde{\rho}}'$ again minimizes the functional $F_\rho[\cdot](\beta,\ell) = F_\beta^\lambda[\cdot](\beta,\ell,0)$, since the $\tilde{\rho}(\lambda_i)$ have minimized the $F_\rho^\lambda[\cdot](\beta,\ell,\lambda)$. Therefore $\underline{\tilde{\rho}}'$ is even a solution of the Thomas-Fermi equation (319), and is unique, and

$$(472) \qquad \underline{\tilde{\rho}}' = \underline{\tilde{\rho}}(0) = \underline{\rho}^{TF} .$$

It remains to show that there exists at all at least one subsequence $\lambda_i \to 0$ such that $\text{s-lim}_{\lambda_i\to 0} \underline{\tilde{\rho}}(\lambda_i)$ exists. As already mentioned in [42] this follows from copying the proof of Lemma (449) where the indices μg disappear and their role is taken by λ. □

Remarks (473): We have made use of the fact that the $\Phi^{\lambda}(N,\beta,\ell,\lambda)$ and therefore also the limit $\Phi^{\lambda}(\beta,\ell,\lambda) = \min_{\rho_\alpha \in \Omega_\alpha} F_{\rho}^{\lambda}[\rho_\alpha](\beta,\ell,\lambda)$ are concave functions in λ. The condition that $\underline{\rho}^{TF}$ is unique is connected to the requirement of absence of phase transitions in the region of thermodynamic parameters under consideration.

Proposition (474): If $\underline{\rho}^{TF}$ is a unique solution of the temperature dependent Thomas-Fermi equation for the canonical ensemble (see (319)), then the following limit exists:

$$(475) \qquad \lim_{N\to\infty} \langle \hat{\rho}_N^{\underline{\alpha}}(f_1,\ldots,f_n)\rangle_\beta = \prod_{k=1}^{n} \int_{\Lambda_\ell} \rho_{\alpha_k}^{TF}(x) f_k(x) d^3x \quad .$$

Proof: The proof follows [42] and uses the Griffiths-Lemma [44] on sequences of concave (convex) functions on the real line in the following form: If $\Phi_N(\lambda)$ is a sequence of concave functions ($N \in \mathbb{N}$), and if $\lim_{N\to\infty} \Phi_N(\lambda) = \Phi(\lambda)$ exists, and if $\Phi(\lambda)$ is differentiable, then $\lim_{N\to\infty} \frac{\partial}{\partial\lambda} \Phi_N(\lambda) = \frac{\partial}{\partial\lambda} \Phi(\lambda)$. By using (404), Lemma (417), Lemma (456), (461), Griffith's Lemma, and Lemma (467) we are led to the result:

$$(476) \qquad \lim_{N\to\infty} \langle \hat{\rho}_N^{\underline{\alpha}}(f_1,\ldots,f_n)\rangle_\beta = \lim_{N\to\infty} \frac{\partial}{\partial\lambda} \Phi^{\lambda}(N,\beta,\ell,\lambda)\big|_{\lambda=0} =$$

$$= \frac{\partial}{\partial\lambda} \Phi^{\lambda}(\beta,\ell,\lambda)\big|_{\lambda=0} = \frac{\partial}{\partial\lambda} F_{\rho}^{\lambda}[\underline{\rho}^{TF}](\beta,\ell,\lambda)\big|_{\lambda=0}$$

$$= \prod_{k=1}^{n} \int_{\Lambda_\ell} \rho_{\alpha_k}^{TF}(x) f_k(x) d^3x. \qquad \Box$$

II.b States on a Hydro-Local C^*-Algebra

We consider equilibrium states for a one component system of purely cosmic matter. To remind how states and observables of infinite quantum systems are usually described we recall

II.ba. The Quasi-Local C*-Algebra

To each bounded open region $\Lambda \subset \mathbb{R}^3$ local observables are represented by the self-adjoint elements of a C*-algebra $\mathcal{A}_\Lambda$. The set $\mathbb{R}^3 = \{\Lambda \subset \mathbb{R}^3 / \Lambda$ an open bounded region$\}$ is directed with respect to the partial ordering of inclusion. One requires the postulate of isotony : If $\Lambda_1 \subset \Lambda_2$ then $\mathcal{A}_{\Lambda_1} \subset \mathcal{A}_{\Lambda_2}$. A C*-algebra $\mathcal{A}$ is given by the C*-inductive limit of $\{\mathcal{A}_\Lambda / \Lambda \in \mathbb{R}^3\}$, written as

$$\mathcal{A} = \overline{\bigcup_{\Lambda \in \mathbb{R}^3} \mathcal{A}_\Lambda} \, . \tag{477}$$

$\mathcal{A}$ is called the quasi-local algebra, containing the quasi-local observables.

States $\omega \in \mathcal{E}(\mathcal{A})$ on the quasi-local algebra are positive, normalized, and linear functionals on $\mathcal{A}$. The states $\omega_\Lambda \in \mathcal{E}(\mathcal{A}_\Lambda)$ converge to $\omega \in \mathcal{E}(\mathcal{A})$ in the weak*-topology

$$\omega = w^* - \lim_{\Lambda \to \mathbb{R}^3} \omega_\Lambda \qquad \text{on } \mathcal{A}, \tag{478}$$

if for each $A \in \mathcal{A}$ and each $\varepsilon > 0$ there exists $\Lambda_o(A,\varepsilon)$ such that for all $\Lambda \supset \Lambda_o(A,\varepsilon)$

$$|\omega_\Lambda(A) - \omega(A)| < \varepsilon \quad . \tag{479}$$

An infinite quantum system for which the usual thermodynamic limit $\Lambda \to \mathbb{R}^3$ exists may be described by a quasi-local C*-algebra. For the thermodynamic Thomas-Fermi limit however a so-called hydro-local C*-algebra (introduced in [45]) may be more suitable.

II.bb. The Hydro-Local C*-Algebra

II.bb.1. Construction of the Algebras

$\mathcal{A}^{(o)}$, C*-algebra of the canonical anticommutation relations (CAR) over $\mathcal{L}_2(\mathbb{R}^3)$,

$\mathcal{A}^{(o)}(\Lambda)$, C*-algebra of the CAR over $\mathcal{L}_2(\Lambda)$, Λ a bounded open region in $\mathbb{R}^3$,

$\tilde{\mathcal{A}}^{(o)}$ consists of polynomials in the fields $\psi(f)$, $\psi^*(g)$ with $f,g \in C_o^2(\mathbb{R}^3)$ and is dense in $\mathcal{A}^{(o)}$.

$\mathcal{A}, \mathcal{A}(\Lambda), \tilde{\mathcal{A}}$ are the gauge invariant subalgebras of $\mathcal{A}^{(o)}, \mathcal{A}^{(o)}(\Lambda), \tilde{\mathcal{A}}^{(o)}$,

$\mathcal{A}_N(\Lambda) = P_N \mathcal{A}(\Lambda) P_N$

with $P_N : \mathcal{H}^{(\Lambda)}_{\text{Fock}} \to \mathcal{H}^{(-)}_{N,\Lambda}$, and P_N is a projection ,

$\mathcal{HL}(\mathcal{A}) = \bigotimes_{x\in\Lambda} \mathcal{A}_x$ with $\mathcal{A}_x = \mathcal{A}$ for all $x \in \Lambda$, be the discrete tensor product of the C*-algebras $\mathcal{A} = \mathcal{A}_x$ $(x \in \Lambda)$ and is called "hydro-local" C*-algebra,

$\mathcal{HL}(\tilde{\mathcal{A}})_o$ is the linear span of the monomials $\bigotimes_{x\in\Lambda} A_x$ with $A_x = A_i \in \tilde{\mathcal{A}}$ if $x = x_i \in \{x_1,\dots,x_k\}$ for some $k \in \mathbb{N}$ and $A_x = 1$ otherwise. $\mathcal{HL}(\tilde{\mathcal{A}})_o$ is dense in $\mathcal{HL}(\mathcal{A})$.

In the Thomas-Fermi scaling an infinity of particles is confined to a bounded region Λ, after performing the thermodynamic limit $N\to\infty$. Therefore its states cannot be locally normal states on the CAR-algebra over $\mathcal{L}_2(\Lambda)$ [46]. The choice of the quasi-local C*-algebra as set of observables is consequently inappropriate, due to the scaling. The reason why the rescaled observables constitute the C*-algebra $\mathcal{HL}(\mathcal{A}) = \bigotimes_{x\in\Lambda} \mathcal{A}_x$ may be understood by the following heuristic arguments [47]:

a) Let us consider an almost infinite Thomas-Fermi system, i.e. $N \gg 1$ is a huge, almost infinite number. Then any arbitrary small region, with volume small compared to $|\Lambda|$, but large compared to $|\Lambda|N^{-1}$, contains almost infinitely many particles and forms already a thermodynamic system.

b) In the scaling used, the volume is fixed, and the Hamiltonian (200) for purely cosmic matter writes by applying the private room argument for fermions and the uncertainty relation

$$\tilde{H}_N \simeq N\cdot N^{-2/3}\,\frac{p^2}{2M} - N^{5/3}\cdot N^{-1}\frac{\kappa M^2}{2\hbar}\,p \quad ,$$

and therefore the momentum per particle p grows proportional to $N^{1/3}$, which means that a particle can be localized in a region of diameter proportional to $N^{-1/3}$. Correlations between particles should therefore exist at a linear range of magnitude $N^{-1/3}$, whereas for larger distances they should tend to zero.

c) Since in a region of diameter $N^{-1/3}$ there are only finitely many particles, the forces between these particles can be neglected. Therefore the correlations should only be determined by the kinetic energy, thus they correspond to those of free particles. The equilibrium state of the infinite Thomas-Fermi scaled system should therefore be of the form $\bigotimes_{x\in\Lambda} \omega_{\rho(x)}$, where at each point $x\in\Lambda$, $\omega_{\rho(x)}$ denotes the equilibrium state of an <u>ideal</u> Fermi gas at the prevailing local density $\rho(x)$. (The ρ is determined by <u>global</u> minimum conditions for the free energy). The algebra of observables of the infinite Thomas-Fermi system should be chosen by attaching to each point $x\in\Lambda$, which represents already an infinite thermodynamic system, a quasi-local C*-algebra $\mathcal{A}$, and forming for the entire system the tensor product $\bigotimes_{x\in\Lambda} \mathcal{A}_x$ of identical copies $\mathcal{A}_x = \mathcal{A}$.

d) The name "hydro-local" for this algebra $\bigotimes_{x\in\Lambda} \mathcal{A}_x$ was chosen, because in hydrodynamics usually infinitesimal volume elements of the fluid are considered, which contain already infinitely many molecules. Those infinitesimal fluid volumes are regarded as thermodynamic systems by themselves, carrying thermodynamic quantities like temperature, density etc.

II.bb.2. Existence of the Limiting Equilibrium State

Scaling and localization of the fields around x is expressed by the map $\sigma : \mathbb{R}_+ \times \Lambda \to \mathrm{Aut}\,(\mathcal{A}^{(o)})$:

$$\sigma(\nu,x)\psi(f) = \psi(\tilde{\sigma}(\nu,x)f), \tag{480}$$

$$(\tilde{\sigma}(\nu,x)f)(y) = \nu^{1/2} f(\nu^{1/3}(y-x)). \tag{481}$$

The restrictions of $\sigma(\nu,x)$ to $\mathcal{A}$ and $\tilde{\mathcal{A}}$ are automorphisms. This automorphism connects the hydro-local C^*-algebra with $\mathcal{A}_N(\Lambda)$ by $J_N: \mathcal{HL}(\tilde{\mathcal{A}})_o \to \mathcal{A}_N(\Lambda)$:

$$(482) \qquad J_N(\bigotimes_{x\in\Lambda} A_x) = \prod_{+\, x\in\Lambda} P_N(\bar{\sigma}(N,x)A_x)P_N$$

where Π_+ signifies the symmetrized product and $(\bar{\sigma}(N,x)A) = (\sigma(N,x)A)$ if $\sigma(N,x)A \in \mathcal{A}(\Lambda)$ and $(\bar{\sigma}(N,x)A) = 0$ otherwise.

Let ω_ρ be the Gibbs state on $\mathcal{A}$ of an infinite ideal Fermi gas with density ρ, and let ω_N be the Gibbs state on $\mathcal{A}_N(\Lambda)$ for the gravitational system, given by the Hamiltonian (200). The following version of the statement

$$(483) \qquad w^* - \lim_{N\to\infty} \omega_N \circ J_N = \bigotimes_{x\in\Lambda} \omega_{\rho^{TF}(x)} \quad \text{on } \mathcal{HL}(\mathcal{A})$$

has been proven in [45]:

<u>Theorem</u> (484) [45]: If ρ^{TF} is a unique solution of the temperature dependent Thomas-Fermi equation for the canonical ensemble (and for purely cosmic matter) and if h is a continuous bounded function on Λ^k for some (arbitrary but finite) $k \in \mathbb{N}$, then the following thermodynamic Thomas-Fermi limit for states exists:

$$(485) \qquad \lim_{N\to\infty} \int_{\Lambda^k} dx_1 \ldots dx_k (\omega_N \circ J_N)(\bigotimes_{i=1}^{k} A_{x_i}) h(x_1,\ldots,x_k)$$

$$= \int_{\Lambda^k} dx_1 \ldots dx_k \prod_{i=1}^{k} \omega_{\rho^{TF}(x_i)} (A_{x_i}) h(x_1,\ldots,x_k)$$

for all $A_{x_1},\ldots,A_{x_k} \in \tilde{\mathcal{A}}$.

<u>Proof</u>: The idea of the proof is similar to Baumgartner's conception ([42], Chapter II.a.) of introducing a generalized free energy, and expressing the state (resp. the correlation functions) by the first derivative of the limiting generalized free energy. Theorem (484) is related to Proposition (474) (for the special case of purely cosmic matter). We shall omit the proof of (484) here and refer to [45].

Exercise (486): To exemplify the convergence of states on the hydro-local algebra, treat the limit of equilibrium states for the grand canonical ensemble of an ideal Fermi gas in an external potential in one dimension. This is a model close to the Thomas-Fermi model for two-body forces. The gas is enclosed in a container $[-L,L]$ on the real line. The single particle Hamiltonian in a Thomas-Fermi-like scaling is

$$(487) \qquad h_L = -\frac{1}{2}\frac{d^2}{dx^2}\bigg|_L + V(\tfrac{x}{L}) - \mu_L ,$$

where the Laplacean has Dirichlet boundary conditions, and μ_L is determined by the requirement that the mean density is fixed to be equal to $\rho > 0$ for all L. The connection with the Thomas-Fermi scaling used in (200) is perhaps best seen by writing the Hamiltonian in the equivalent scaling

$$(488) \qquad h_L' = -\frac{1}{2}L^{-2}\frac{d^2}{dx^2} + V(x) - \mu_L',$$

on the box $[-1,1]$ of fixed volume. Note also that the expression (488) indicates that the Thomas-Fermi thermodynamic limit $L \to \infty$ is connected with a kind of classical limit with formally $\hbar \sim L^{-1} \to 0$. Find conditions on V, such that h_L is essentially self-adjoint on a suitable domain in $\mathcal{L}_2(\mathbb{R})$, h_L has a discrete spectrum only, and $\exp(-\beta h_L)$ is trace-class. Consider the Fermi field algebra $\mathcal{A}^{(o)}$ represented on the Fermi Fock space with creation and annihilation operators $a(f)$ and $a^+(f)$ for appropriate test functions. Prove that the chemical potentials converge in the limit $L \to \infty$ to some μ. Note that the system is quasi-free and the equilibrium state is uniquely given by the two-point function. Let ω_L denote the Gibbs state for inverse temperature β on $\mathcal{A}$ and consider the following automorphism on $\mathcal{A}^{(o)}$

$$(489) \qquad \gamma(L,q)a^{\#}(f) = a^{\#}(\tilde{\gamma}(L,q)f)$$

$$(490) \qquad (\tilde{\gamma}(L,q)f)(x) = f(x-qL).$$

Prove that for all $f \in \mathcal{D}(\mathbb{R})$

(491) $$\lim_{L\to\infty} \omega_L(\gamma(L,q)a^+(f)a(f)) =$$

$$= \omega_q(a^+(f)a(f)) = \int \frac{dk}{2\pi} \frac{|\tilde{f}(k)|^2}{1+e^{\beta(\frac{1}{2}k^2+V(q)-\mu)}} ,$$

where $\tilde{f}$ is the Fourier transform of f. Define a map $J_L: \mathcal{HL}(\tilde{\mathcal{A}})_o \to \mathcal{A}$ by

(492) $$J_L(\underset{q\in\mathbb{R}}{\otimes} A_q) = \prod_{q\in\mathbb{R}} \gamma(L,q)A_q$$

Show that for arbitrary $n \geq 1$, $m \geq 1$:

(493) $$\lim_{L\to\infty} \omega_L(a^+(f^L_{q_1})\dots a^+(f^L_{q_n})a(g^L_{q_m})\dots a(g^L_{q_1}))$$

$$= \delta_{nm} \prod_{i=1}^{n} \omega_{q_i}(a^+(f)a(g))$$

with, for $f \in \mathcal{D}(\mathbb{R})$,

(494) $$f^L_q = \tilde{\gamma}(L,q)f.$$

Prove that on $\mathcal{HL}(\mathcal{A})$

(495) $$w^* - \lim_{L\to\infty} \omega_L \circ J_L = \underset{q\in\mathbb{R}}{\otimes} \omega_q.$$

Note that due to the scaling, in the limit $L \to \infty$ the functions $f^L_{q_1}$ and $f^L_{q_2}$, where $f \in \mathcal{D}(\mathbb{R})$, become orthogonal if $q_1 \neq q_2$. For a treatment of similar problems (for Bose statistics) see [48] and [82].

III. *EXISTENCE OF PHASE TRANSITIONS*

In the evolution of stars and galaxies instabilities are observed - the most impressive of which is the supernova phenomenon - which are, compared to the lifetime of stars, rapid changes of the state. Some spectacular evolutionary mutabilities like the gravitational collapse towards a black hole are of relativistic or at least semirelativistic origin; whereas the creation of a neutron star in the centre of a celestial body undergoing a supernova transition is commonly regarded as an effect of nonrelativistic quantum mechanics and Newtonian gravity. It is therefore of interest to investigate whether model-systems of purely gravitating fermions exhibit the phenomenon of phase transition by themselves. More precisely the question ought to be posed as asking for the power of the set of extremal equilibrium states, which are those extremal KMS-states, which minimize the free energy functional, which is not implicit because of the Thomas-Fermi scaling. If the solution of the temperature dependent Thomas-Fermi equation is unique, then it follows from Theorem (484) that the limit of the equilibrium state is already extremal. If non-uniqueness of the solutions of the Thomas-Fermi equation can be shown, then we expect the local Gibbs state to converge to a convex combination of extremal states, which are the Gibbs states $\omega_{\rho^{TF}(x)}$ of those ideal Fermi gases, which possess the corresponding solutions $\rho^{TF}(x)$ as densities. This connection ought to be presumed, however, it is not yet proven. We shall therefore demonstrate that non-uniqueness leads also to non-analyticity and thus fits into the definition of a phase transition according to Ehrenfest.

III.a. Non-Uniqueness of Solutions of the Temperature Dependent Thomas-Fermi Equation

The aim of this section is to outline an analytical proof for the existence of a phase transition in purely cosmic matter. We shall also discuss the absence of such a phase transition in the Thomas-Fermi theory of ordinary matter. We start with mentioning some historical landmarks in the endeavour of discerning the nature and deriving the existence of the gravitational phase transition. There is a close similarity to the kind of phase transition we envisage in hadron physics, e.g. between quark matter and hadrons. We also sketch some breakthrough in the history of this subject.

III.a.1. Brief Historical Surveys

1.1) Stability of Matter

The early history of stability considerations is carefully reviewed in [7]. The more recent developments are already stated in Chapter I.a.

1.2) Phase Transitions in Cosmic Matter

<1962 Gravitational instabilities and existence of negative specific heat has been predicted by astrophysicists on the basis of the classical model of stars [49,24].

1962 Antonov pointed out two properties of the entropy in a classical system of gravitating particles [50]:
First result: Only spherical symmetric systems can have a local maximum of the entropy for given energy.
Second result: In a classical system of gravitating particles there is no global maximum of the entropy.

1968 Lynden-Bell [51] and Lynden-Bell and Wood [52] discussed Antonov's result. They showed the existence of negative specific heat which leads to a gravothermal catastrophe (in classical systems): The density becomes singular in the origin, no equilibrium states exist for large radii. They indicated the association of phase transitions with the gravothermal catastrophe when the classical potential is modified or Fermi statistics is used.

1970 Thirring [53] rederived independently Antonov's and Lynden-Bell's analysis. He also showed that if the potential is modified in the origin and the system is enclosed in a box, equilibrium states can exist for all radii. He was first realizing (for classical systems) that the region of negative specific heat is bridged by a phase transition in the canonical ensemble.

1971 Hertel and Thirring [39] demonstrated the existence of a phase transition in the canonical ensemble of a system of gravitating fermions by solving numerically the Thomas-Fermi temperature dependent equation.

1972 Aronson and Hansen [54] supported Lynden-Bell's and Wood's [52] and Thirring's [53] conclusion that equilibrium states do always exist if the potential is modified in the origin. They showed on the basis of computer calculations that a system of classical gravitating hard spheres should possess a phase transition in the canonical ensemble.

1977 Hertel [55] reported on an improvement of the numerical calculations [39].

1980 Pflug [56] proved analytically the existence of a phase transition for gravitating fermions, if the thermodynamics is simplified by performing the infinite volume limit for suitably volume-rescaled observables of the Thomas-Fermi system.

1.3) Phase Transitions in Hadronic Matter

Analogously to the behaviour of gravitating fermions negative specific heat occur also in elementary particle physics. Although there is an intrinsic resemblance between both theories, not yet completely elaborated, we shall only cursory touch the subject and dispense with quoting all references. We describe in a few sentences some historical steps [57] of the progress in understanding the phase transition in hadronic matter. As entrance gate to the field we recommend the article [58] by Çelik and Satz.

In 1964 Gell-Mann and Zweig postulated the quark infra-structure of hadrons. One year later Hagedorn [59] showed that an exponential resonance spectrum $M^a e^{bM}$ implies the existence of an ultimate temperature $T_H = 1/b$. Since 1969 various authors (Ivanenko and Kurdgelaidse, Itoh, Baym and Chin, Collins and Perry, Chapline and Nauenberg, and others) discussed hadron-quark, resp. nuclear matter - hadron matter phase transitions in the interior of stars, in heavy ion collisions, and in elementary particle reactions. It was in 1972 that Carlitz [60], and in 1974 that Cabibbo and Parisi [61] realized that $T_c = 1/b$ represents a critical temperature for a second order phase transition, and that the canonical partition function for hadronic matter with the level density $M^a e^{bM}$ diverges by surpassing T_c. If $a \geq -7/2$ the energy density also diverges at T_c, which then constitutes an ultimate temperature of the system, whereas for $a < -7/2$ the energy density remains finite and phase transitions become possible. Furthermore Carlitz showed that the specific heat becomes negative in the microcanonical ensemble. Since 1978 extensive investigations started to explore the hadron - quark matter phase transition both in phenomenology (bootstrap model, bag model, van der Waals approach), as well as in the framework of quantum chromodynamics, studying, e.g., perturbation theory or instanton effects [62].

III.a.2. Uniqueness for Atoms and Ordinary Matter

We have already briefly described the ground-state Thomas-Fermi theory for ordinary matter. In this case a solution of the Thomas-Fermi equation exists and is unique (Theorem (32)). Now we consider the temperature dependent Thomas-Fermi theory for ordinary matter and demonstrate again uniqueness of the solution. In the model used, ordinary matter is composed of N electrons of mass 1/2 and charge -1 in Coulomb interaction with K infinitely heavy nuclei with charges $z_1,\dots,z_K$. The Hamiltonian of this system of ordinary matter is given formally by the equations (21-23). We assume that strict neutrality $\sum_{j=1}^{K} z_j = N$ holds for this one component system of normal matter. Atoms can be described in this framework. They are special cases of ordinary matter, obtained from (21-23) when $K = 1$, $z_1 = Z$, $R_1 = 0$, $U(\{z_1,R_1\}) = 0$ (and $N = Z$).

The temperature dependent Thomas-Fermi theory for this model of ordinary matter was developed in [63] and in [15] by Narnhofer und Thirring. They showed that the thermodynamic Thomas-Fermi limit for the grand canonical pressure exists and the temperature dependent Thomas-Fermi equation becomes exact in the limit. The use of coherent states is the key for understanding this derivation of the Thomas-Fermi theory and it also elucidates the character of a classical limit for the thermodynamic limiting in the Thomas-Fermi scaling. The application of entropy inequalities with coherent states in [63,15] simplifies the construction of the Thomas-Fermi limit considerably, compared to our presentation in Section I.bb. However, this technique fails in the presence of gravitation. Concerning the question of uniqueness and existence of solutions, we investigate briefly the following temperature dependent Thomas-Fermi equation for ordinary matter in the canonical ensemble

$$\rho(x) = q \int \frac{d^3p}{(2\pi)^3} \left(1 + e^{\beta(p^2+W(x)-\mu)}\right)^{-1} \quad , \tag{496}$$

with

$$(497)\qquad W(x) = \int\limits_{\Lambda} \frac{\rho(y)}{|x-y|}\, d^3y - \sum_{j=1}^{K} \frac{z_j}{|x-R_j|} \quad ,$$

$$(498)\qquad \int\limits_{\Lambda} \rho(x)\, d^3x = n \quad ,$$

(499) ρ minimizes the free energy functional.

Since in Section I.bb. or in [63,15] there is an explicit construction of the limit points, the question of existence of solutions of the above Thomas-Fermi equation is already proved:

Lemma (500): For each $\beta > 0$ and each bounded, connected and measurable subset $\Lambda \subset \mathbb{R}^3$ there exists at least one solution of the temperature dependent Thomas-Fermi equation (496-499).

Uniqueness of the solutions follows from the next two lemmata:

Lemma (501): For each $\beta > 0$ and each radius R of a ball $\Lambda = \Lambda_R \subset \mathbb{R}^3$ there exists at most one solution of the temperature dependent Thomas-Fermi equation (496-499) for atoms.

Proof: We follow [39, Ref.13]. Assuming spherical symmetric boundary conditions, the potential

$$(502)\qquad U(|x|) = \int \frac{\rho(|y|)}{|x-y|}\, d^3y$$

generated by the electrons obeys Poisson's equation

$$(503)\qquad \Delta U = -4\pi\rho \quad ,$$

which simplifies in this case to

$$(504)\qquad \frac{1}{r^2}\frac{\partial}{\partial r}\left(r^2\,\frac{\partial U(r)}{\partial r}\right) = -4\pi\rho(r) \quad , \quad r = |x| \ .$$

Defining the new variable

(505) $\quad r^{-1}\Phi(r) = U(r) - \mu,$

leads with (504) and (496) for atoms to

(506) $$\Phi''(r) = -C\, r \int_0^\infty d\varepsilon\sqrt{\varepsilon}\, (1 + e^{\beta(\varepsilon + r^{-1}(\Phi(r)-Z)})^{-1}$$

with $\Phi(0) = 0$, $C > 0$ some constant, and the prime denotes differentiation with respect to r. If we assume the existence of two solutions Φ_1 and Φ_2 with $\Phi_1'(0) = U(0) - \mu = \lambda = \Phi_2'(0)$, then $\Phi_1(r) = \Phi_2(r)$ for all r follows from the uniqueness of solutions of the "λ-equation" in Lemma (532), already proven in [39]. Let us therefore suppose that $\Phi_1'(0) > \Phi_2'(0)$. By continuity and after integration, $\Phi_1'(r) \geq \Phi_2'(r)$ and $\Phi_1(r) \geq \Phi_2(r)$ for all r in a sufficiently small neighbourhood of 0. From (506) we observe, that Φ'' is monotonically increasing in Φ. Therefore $\Phi_1(r) \geq \Phi_2(r)$ holds for all $0 \leq r \leq R$, which implies $\rho_1(r) \leq \rho_2(r)$, contradicting the normalization condition (498). This reasoning does not hold in case of gravitation, because then Φ'' becomes monotonically decreasing in Φ. □

Lemma (507): For rather general domains $\Lambda \subset \mathbb{R}^3$, the solution of the temperature dependent Thomas-Fermi equation for ordinary matter in the grand canonical ensemble is unique.

Proof: This generalization of Lemma (501) was proved in [63,15]. The thermodynamic Thomas-Fermi limit for the grand canonical pressure turns out to be

(508) $$p = \inf_{\substack{\rho \in \mathcal{L}_{5/3}(\Lambda) \\ \rho \geq 0}} p_\rho[\rho]$$

for some pressure functional $p_\rho[\cdot]$, defined in [63,15]. If $\mathcal{L}_0^{(+)}$ is the Hilbert cone of positive measurable functions from $\mathbb{R}^3 \to \mathbb{R}_+$ equipped with the norm

(509) $$\|\rho\|_c^2 = (\rho,\rho)_c = \int d^3x d^3y \, \frac{\rho(x)\rho(y)}{|x-y|} ,$$

then the map $\rho \in \mathcal{L}_0^{(+)} \to p_\rho[\rho] \in \mathbb{R}_+$ has the following properties:

(510) It is weakly lower semicontinuous.

(511) It is strictly convex.

(512) $p_\rho[\rho] \geq \frac{1}{2} \|\rho\|_c^2$.

From this list we can conclude that the infimum is in fact a minimum (and a solution of the Thomas-Fermi equation for the grand canonical ensemble), this minimum belongs to some weakly compact region $\|\rho\|_c^2 < C$, and the minimum is unique. For the details of the proof we refer to [63,15]. □

For further studies of the temperature dependent Thomas-Fermi theory for ordinary matter we recommend the detailed presentation in [15, Vol.4], whereas we shall now turn our attention to cosmic matter.

III.a.3. Non-Uniqueness for Purely Cosmic Matter

The gravitational phase transition should be visible in the canonical and grand canonical ensembles, it is replaced in the microcanonical ensemble by a region of negative specific heat. To demonstrate the appearance of the phase transition in the Thomas-Fermi equation, we shall at best choose the canonical ensemble, which is most naturally tailored for Thomas-Fermi theory and in which Thomas-Fermi theory is also best understood. In addition spherical symmetric boundary conditions are chosen. In starting with the Hamiltonian for purely cosmic matter in the appropriate scaling, we are led to the Thomas-Fermi equation, which describes the infinite system of gravitating fermions exactly. At next we shall cut this complicated equation into three constituents, the "n-equation", the "λ-equation", and the "μ-equation". The normalization constant n can be interpreted as a particle number (although we are considering an infinite system), and λ is a renormalized chemical potential, obtained by shifting the negative chemical potential $-\mu$ by a constant. After considering existence and possible uniqueness of the solutions of the three constituting equations, we study the function $n(\lambda)$, which describes the mass distribution in terms of the renormalized chemical potential. For technical reasons we have actually to investigate the behaviour of the temperature $\beta^{-1}(\lambda)$, but this complication does not matter. The functions $n(\lambda)$ and $\beta^{-1}(\lambda)$ are analytic and non-monotonic. Therefore, there are fixed $n = n(\lambda)$ or fixed $\beta^{-1} = \beta^{-1}(\lambda)$ where we obtain at least two different λ's, yielding the same n or β^{-1} resp. This leads finally to the non-uniqueness of the solutions of the Thomas-Fermi equation.

We do not intend to present every detail of the proofs, since the original papers [64,65] are soon available. It may however be helpful to the interested reader to find a short survey and introduction here, describing the main ideas.

Definition (513): Given $\beta > 0$ and the ball Λ_R with radius R, and $n > 0$. Let

$$(514) \qquad g(\beta,y) := q \int \frac{d^3p}{(2\pi)^3} \left(1 + e^{\beta(p^2+y)}\right)^{-1}$$

hold for every arbitrary $y \in \mathbb{R}$.

The following self-consistency equation is called n-equation:

$$(515) \qquad \rho(x) = T[\rho](x) = g(\beta, W[\rho](x) - \tilde{\mu}[\rho]) \ ,$$

$$(516) \qquad \int_{\Lambda_R} T[\rho](x) d^3x = n \ .$$

Since the function $g(\beta,\cdot)$ is monotonic and continuous, the functional $\tilde{\mu}$ is uniquely determined with (515) by the condition (516).

The potential W for purely cosmic matter is

$$(517) \qquad W[\rho](x) = - \int_{\Lambda_R} \frac{\rho(y)}{|x - y|} \ .$$

Definition (518): Given $\beta > 0$, $R > 0$, and $\mu \in \mathbb{R}$.

The following self-consistency equation is called μ-equation:

$$\rho(x) = T_\mu[\rho](x) = g(\beta, W[\rho](x) - \mu). \tag{519}$$

Definition (520): Given $\beta > 0$, $R > 0$, and $\lambda \in \mathbb{R}$.

The following self-consistency equation is called λ-equation:

$$\rho(x) = G_\lambda[\rho](x) = g(\beta, w[\rho](x) + \lambda), \tag{521}$$

with

$$w[\rho](x) = W[\rho](x) - W[\rho](0). \tag{522}$$

Remark (523): The n-equation can also be equivalently defined by the two conditions: Given $\beta > 0$, $R > 0$ and $n > 0$, then

$$\rho(x) = G[\rho](x) = g(\beta, w[\rho](x) + \tilde{\lambda}[\rho]), \tag{524}$$

$$\int_{\Lambda_R} G[\rho](x)\,d^3x = n, \tag{525}$$

where the functional $\tilde{\lambda}$ is again uniquely defined by condition (525).

Remark (526): The Thomas-Fermi equation (see (319)) is defined by the following conditions:

ρ^0 is a solution of the temperature dependent Thomas-Fermi equation if and only if:

$$\rho^0 \text{ is solution of the n-equation}, \tag{527}$$

$$\Phi = \min_\rho F_\rho[\rho] = F_\rho[\rho^0]. \tag{528}$$

The free energy functional F_ρ is given by (312), specialized for the one component system of purely cosmic matter, with normalization to n instead of 1, and with μ replaced by the functional $\tilde{\mu}$.

Lemma (529): For each $\beta > 0$, $R > 0$, and $n > 0$ there exists at least one solution of the n-equation.

Proof: From the convergence of the free energy in the thermodynamic Thomas-Fermi limit to the minimum of the free energy functional the existence of a solution of the Thomas-Fermi equation follows immediately from the existence of the limit point. This proves also the existence of solutions of the n-equation as a special case. However, we shall give an existence proof here, being independent of the construction of the Thomas-Fermi limit for the free energy: The possible solutions of the fixed point problem (515,516) are elements of a set of integrable, equicontinuous and bounded functions. Endowed with the $\mathcal{L}_\infty$-topology this set is compact and convex by Ascoli's theorem and the lemma of Heine and Borel. We also observe that the functional T maps this set of possible solutions into itself, and T is continuous in the $\mathcal{L}_\infty$-topology. The fix point theorem of Schauder and Tychonoff can therefore be applied and yields the existence of at least one fixed point. □

Remark (530): Rewriting the μ-equation as Poisson's differential equation leads to a boundary value problem. The corresponding initial value problem is the λ-equation. Therefore it is easier to work with the λ-equation than with the μ-equation. The λ-equation shows uniqueness of the solution (Lemma (532)). This cannot be expected from the μ-equation. Indeed one can show [64] that only for sufficiently small μ the μ-equation exhibits uniqueness of the solution, whereas by relaxing this condition there is only existence of solutions, and if $\mu > \mu_o$, μ_o implicitly defined by

$$(531) \qquad g(\beta,-\mu_o) = \frac{243}{16}\pi R^{-6}(1 + e^{-\beta\mu_o})^2 \quad ,$$

there exists no solution of the μ-equation.

Lemma (532): For each $\beta > 0$, $R > 0$, and $\lambda \in \mathbb{R}$ there exists exactly one (unique) solution of the λ-equation.

Proof: This lemma has first been proved in [39] by considering the λ-equation as equivalent to Poisson's differential equation with a non-linear source term and applying standard theorems for ordinary differential equations with initial values. In [55] it has been observed that the λ-equation is equivalent to the following integral equation

(533) $$K_\lambda[w](r) := \int_0^r s^{-2}ds \int_0^s t^2 dt\, 4\pi g(\beta, w(t)+\lambda) ,$$

(534) $$K_\lambda[w](r) = w(r).$$

The second proof of Lemma (532), given in [55], is based on the metric

(535) $$d^\alpha(u,v) = \sup_{0\leq r\leq R} e^{-r/\alpha}|u(r) - v(r)| ,$$

which turns the set $\Omega_\lambda = \{w : [0,R] \rightarrow \mathbb{R}_+$, continuous, and $0 \leq w(r) \leq$ $\leq (4\pi/6)R^2\, g(\beta,\lambda)\}$, containing all possible solutions of the λ-equation, into a complete topological space. With respect to the metric (535), and for sufficiently large α, the functional K_λ is easily seen to be a contraction map on Ω_λ and therefore Banach's fix point theorem concludes the proof. □

Our final aim is to investigate uniqueness or non-uniqueness of the solutions of the Thomas-Fermi equation, composed of the λ- and n-equation. First we find that in the high temperature domain, when the thermal energy $\frac{3}{2}$ kT is large compared to the gravitational binding energy, the solution is unique. At second we show that uniqueness cannot hold for every choice of the thermodynamic parameters (β,R,n), which leads to the existence of a phase transition.

Lemma (536): Let $\hat{g}$ be defined by the right-hand side of (426) for $\lambda = 0$. The set

(537) $$\mathcal{N} = \{ (\beta,R,n) \in \mathbb{R}_+^3 / \hat{g}\kappa M^2 \beta 2\pi R^2 < \tfrac{1}{2} \}$$

has the following properties:

(538) $\mathcal{N}$ is not empty.

(539) For each element of $\mathcal{N}$ there exists exactly one (unique) solution of the temperature dependent Thomas-Fermi equation for purely cosmic matter.

(540) If $(\beta,R,n)\in\mathcal{N}$ then $\frac{2}{3}\,\beta\,\frac{\kappa M^2 n}{R} < \frac{2}{9}$.

Proof: We give only the idea of the proof which leads to uniqueness. For the detailed argumentation we refer to [64]. To every $\lambda\in\mathbb{R}$ there is a unique solution ρ_λ of the λ-equation and consequently a well-defined function

$$n(\lambda) = \int_{\Lambda_R} d^3x\rho_\lambda(x) \quad . \tag{541}$$

Furthermore there is a λ_o, such that $n(\lambda)$ is strictly monotonically decreasing for all $\lambda > \lambda_o$ (see Lemma (543)). Assume that to some $(\beta,R,n)\in\mathcal{N}$ there are two solutions ρ_1 and ρ_2 of the Thomas-Fermi equation. Since from the self-consistency equation the possible λ's are bounded, i.e., there exist λ^ℓ, λ^u such that

$$\lambda^\ell \leq \tilde{\lambda}[\rho] \leq \lambda^u \tag{542}$$

we have $\lambda_1 = \tilde{\lambda}[\rho_1] \geq \lambda^\ell$ and $\lambda_2 = \tilde{\lambda}[\rho_2] \geq \lambda^\ell$. But the set $\mathcal{N}$ is just constructed in a way such that $\lambda^\ell > \lambda_o$, and therefore by strict monotonicity, the equation $n = n(\lambda_i)$ has the unique solution $\lambda_1 = \lambda_2$, and with uniqueness of the solutions of the λ-equation we finally conclude $\rho_1 = \rho_{\lambda_1} = \rho_{\lambda_2} = \rho_2$. □

Lemma (543): Let ρ_λ be the solution of the λ-equation for given $\lambda\in\mathbb{R}$. Then the function $\lambda \to n(\lambda) = \int_{\Lambda_R} \rho_\lambda(x)d^3x$ has the following properties:

(544) To each $\beta > 0$, $R > 0$ there exists λ_o such that for all $\lambda > \lambda_o$: $n(\lambda)$ is strictly monotonically decreasing.

$$\lim_{\lambda\to-\infty} n(\lambda) = \infty \quad . \tag{545}$$

(546) $n(\lambda) \leq |\Lambda_R| g(\beta,\lambda)$.

(547) There exist a $\beta_o > 0$ and a $R_o > 0$ and a neighbourhood $\mathcal{U}(\lambda_o')$ of some λ_o' such that $n(\lambda)$ is strictly monotonically increasing on $\mathcal{U}(\lambda_o')$.

(548) $\lambda \to n(\lambda)$ is real analytic.

Proof: Part (546) is trivial, for part (545) we refer to [64]. The proof of analyticity (548) relies on the possibility to write Poisson's differential equation for our case as a system of holomorphic differential equations. More details can be found in [64].

Strict monotonicity (544) for large λ follows from the inequality

$$(549) \qquad \| w[\rho_{\lambda_1}] - w[\rho_{\lambda_2}] \|_\infty < |\lambda_1 - \lambda_2| ,$$

which is a consequence of (522,517,521) for $2\pi R^2 \beta g(\beta,\lambda_2) < 1/2$. This implies $n(\lambda_1) < n(\lambda_2)$ if $\lambda_1 > \lambda_2$. Finally choose $2\pi R^2 \beta g(\beta,\lambda_o) = 1/2$.

Non-monotonicity (547) of $n(\lambda)$ was proved in [65]. The λ-equation for the renormalized potential $w(r,\lambda)$ written as differential equation can be reduced to a one dimensional Schrödinger equation to eigenvalue zero and for the wave function $f(r) = r(1 + \dot{w}(r,\lambda))$, where the dot denotes differentiation with respect to λ. By utilizing Sturm's theorem, one can show that the solution $f(r)$ becomes negative for some r and λ. On the other hand, if $n(\lambda)$ would be everywhere monotonically decreasing, then it would follow that $(1 + \dot{w}(r,\lambda)) \geq 0$ for all r and λ. For the details see [65]. □

Theorem (550): There exist $\beta > 0$, $R > 0$, and $n > 0$ such that the temperature dependent Thomas-Fermi equation for purely cosmic matter has at least two solutions.

Proof:

1st step: Properties of $\beta^{-1}(\nu)$ with $\nu = \beta\lambda$.

Since we do not know the continuity of the free energy as function of n, we have to concentrate the analysis on the variable β. From the properties of $n(\nu)$, we infer the analogous properties of $\beta : (R,n,\nu) \in \mathbb{R}_+^2 \times \mathbb{R} \to \beta(\nu) \in \mathbb{R}_+$. The function $\beta(\nu)$ is implicitly defined by

$$(551) \qquad n = \beta(\nu)^{-3/4} \int_0^{\beta(\nu)^{-1/4}R} 4\pi\xi^2 \, g(1,\xi^{-1}v(\xi)+\nu)d\xi,$$

where $v(\xi)$ is the solution of the λ- (or better ν-) equation:

$$(552) \qquad v''(\xi) = \xi \frac{q}{\pi} \int_0^\infty d\eta\sqrt{\eta}(1 + e^{\eta+\xi^{-1}v(\xi)+\nu})^{-1}$$

with $v(0) = v'(0) = 0$, and v' denotes differentiation with respect to ξ. Comparing the two functions $\beta^{-1}(\nu)$ and $n(\nu)$ leads to the following results:

(553) $\nu \to \beta^{-1}(\nu)$ is real analytic.

(554) $\lim_{\nu\to\infty} \beta(\nu) = 0$.

(555) $\nu \to \beta^{-1}(\nu)$ is non-monotonic .

<u>2nd step</u>: <u>Continuity of the Free Energy in β</u>.

Since the limiting free energy is the limit of ordinary, but rescaled free energies, the limit function $\beta \to \beta\Phi_{W_{(n)}}(n,\beta,R)$ shares the property of concavity. Therefore $\Phi_{W_{(n)}}(n,\beta,R)$ is continuous in $\beta \in (0,\infty)$.

<u>3rd step</u>: <u>Strong Continuity of the Free Energy Functional</u>.

The functional $\rho \to F_\rho[\rho](n,\beta,R)$ is strongly continuous with respect to the $\mathcal{L}_\infty$-topology for given R, n, and uniformly for all $\beta \in B \subset (0,\infty)$. More precisely: If for a sequence $\{\beta_i\}$, $i \in \mathbb{N}$ with $\lim_{i\to\infty} \beta_i = \beta_0$ the strong limit $\text{s-}\lim_{i\to\infty} \rho(\beta_i) = \rho_0$ exists, then

$$(556) \qquad \lim_{i\to\infty} F_\rho[\rho(\beta_i)](n,\beta_i,R) = F_\rho[\rho_0](n,\beta_0,R).$$

4th step: Convergence of the Densities.

The strong convergence of the densities $\rho(\beta_i)$, postulated in the third step, holds for the solutions of the ν-equation (or λ-equation). Let ρ_{ν_i,β_i} denote the solution of the λ-equation for given β_i and ν_i. Choose the sequence β_i such that $\lim_{i\to\infty} \beta_i = \beta_o$, and associate with every solution of the n-equation $\rho(\beta_i)$ a $\nu_i = \beta_i\tilde{\lambda}[\rho(\beta_i)]$. If $\lim_{i\to\infty} \nu_i = \nu_o$, then

(557) $$\text{s-}\lim_{i\to\infty} \rho_{\nu_i,\beta_i} = \rho_{\nu_o,\beta_o} ,$$

and ρ_{ν_o,β_o} is the solution of the λ-equation for given ν_o and β_o.

5th step: $\nu_{TF}(\beta)$ is Discontinuous.

Let us assume from now on:

(558) For every (β,R,n) the solution ρ^{TF} of the Thomas-Fermi equation is unique.

A consequence of this uniqueness assumption is that $\beta \to \nu_{TF}(\beta) := \beta\tilde{\lambda}[\rho^{TF}(\beta)]$ is a well-defined function. However the graph $\beta^{-1}(\nu)$ shows non-monotonicity, i.e., there is a region of β's, where the equation $\beta^{-1}(\nu_j) = \beta^{-1}$ has at least three solutions. Therefore the $\nu_{TF}(\beta)$-function cannot be continuous, but has to jump. Consequently there is a β_o and subsequences β_{i_j} and β_{i_k} both converging to β_o, but

(559) $$\lim_{i_j\to\infty} \nu_{TF}(\beta_{i_j}) = \nu_1 \neq \nu_2 = \lim_{i_k\to\infty} \nu_{TF}(\beta_{i_k}) .$$

6th step: The Minimum of the Free Energy Functional

Since $\rho^{TF} = \rho_{\nu_{TF}}$ is the solution of the Thomas-Fermi equation, assumed to be unique, it minimizes the free energy functional $F_\rho[\cdot]$ yielding the limiting free energy $\Phi_{W_{(n)}}$. Now consider the sequence β_i converging to the β_o of the fifth step and its two subsequences. Then

(560) $$\lim_{i_j\to\infty} F_\rho[\rho_{\nu_{TF}(\beta_{i_j}),\beta_{i_j}}](\beta_{i_j}) = F_\rho[\text{s-}\lim_{i_j\to\infty} \rho_{\nu_{TF}(\beta_{i_j}),\beta_{i_j}}](\beta_o)$$

$$= F_\rho[\rho_{\nu_1,\beta_o}](\beta_o),$$

and analogously

(561) $$\lim_{i_k\to\infty} F_\rho[\rho_{\nu_{TF}(\beta_{i_k}),\beta_{i_k}}](\beta_{i_k}) = F_\rho[\rho_{\nu_2,\beta_o}](\beta_o).$$

By continuity of the free energy we have

(562) $$F_\rho[\rho_{\nu_1,\beta_o}](\beta_o) = \lim_{i_j\to\infty} \Phi_{W_{(n)}}(n,\beta_{i_j},R) = \Phi_{W_{(n)}}(n,\beta_o,R) =$$

$$= \lim_{i_k\to\infty} \Phi_{W_{(n)}}(n,\beta_{i_k},R) = F_\rho[\rho_{\nu_2,\beta_o}](\beta_o).$$

However, since with $\nu_1 \neq \nu_2$ also $\rho_{\nu_1,\beta_o} \neq \rho_{\nu_2,\beta_o}$ is valid, (562) contradicts (558).
For all details and proofs we refer to [64]. □

Corollary (563): There exist $\beta > 0$, $R > 0$, and $n > 0$ for which an infinite system of gravitating fermions shows an Ehrenfest phase transition of the first kind.

Proof: If ρ_{TF} is unique, then the first derivative

(564) $$\tilde{\mu}[\rho_{TF}] = \frac{\partial}{\partial n} \Phi_{W_{(n)}}(n,\beta,R)$$

exists [39]. For two solutions $\rho_{TF,1}$ and $\rho_{TF,2}$ on the one hand we stay at the same free energy $\Phi_{W_{(n)}} = F_\rho[\rho_{TF,1}] = F_\rho[\rho_{TF,2}]$, but on the other hand the chemical potentials $\tilde{\mu}[\rho_{TF,1}]$ and $\tilde{\mu}[\rho_{TF,2}]$ are different as we saw from the proof of Theorem (550). Therefore the above first derivative cannot exist at the non-uniqueness point. □

Remarks (565): Very similarly the first derivative

$$(566) \qquad P[\rho] = -\frac{\partial}{\partial V}\Phi_{W_{(n)}}(n,\beta,R) ,$$

with $V = \frac{4\pi}{3}R^3$, does not exist (or the right- and left-hand side derivatives don't coincide) at the phase transition point. The mechanical pressure

$$(567) \qquad P[\rho] = \frac{2}{3}q\int \frac{d^3p}{(2\pi)^3}\, p^2 \,(1 + e^{\beta(p^2 - \frac{n}{R} - \tilde{\mu}[\rho])})^{-1}$$

becomes discontinuous, describing an implosion or explosion respectively.

The point (β_o, R_o, n_o) of the occurence of the phase transition stretches to a line, because of the scaling properties. The transition shows up at every point $(\gamma^{-4/3}\beta_o,\ \gamma^{-1/3}R_o,\ \gamma n_o)$ for arbitrary $\gamma > 0$.

III.b. Applications to Astrophysics

III.b.1. Applicability of the Equilibrium Model

Our model cannot directly be applied to stars, because:

a) they are not isothermal,
b) they have inhomogeneous chemical composition,
c) they have internal (nuclear) energy sources (and weak interaction processes),
d) the time scales do not allow sufficient relaxation. The evolution is not always quasistatic,
e) the stars cannot be enclosed in rigid containers, etc.

There are some few cases for a realistic application to astrophysics:

III.b.2. Creation of Stars

We give the following interpretation [66]:

When an interstellar HI-cloud of approximately 100 K and 100 atoms/cm^3 reaches the Jeans-instability (mass > 10^4 sun masses) the consecutive contraction leads to ionisation of the HI and to an expansion of the HII-gas. This involves the creation of globules because of hydrodynamic instabilities. Diameter and density of globules are 10^{12} - 10^{13} km, resp. 10^4 - 10^5 atoms/cm^3. They are the statistical system for a theory of gravitating fermions for $V \simeq \infty$ (resp. $\rho \simeq 0$), and represent the non-condensed phase. A theory of gravitating particles in such an infinite configuration space has been developed by A. Pflug [56]. The graph of the (renormalized) entropy at $V \simeq \infty$ versus the energy shows a phase transition (in the microcanonical ensemble). The transition marks the birth and the death of a star, whereas during his lifetime he passes non-equilibrium states which are not on the equilibrium graph.

III.b.3. Galaxies and Star Clusters

In galaxies the stars are the particles and the dense centre may represent the condensed phase. The time scale governing this phase transition should be slow [53]. Although the relaxation times are too long there might be a "violent relaxation" [67] during the galaxies birth stages. If in galactic nuclei the star-star relaxation time is short enough it might undergo a gravothermal catastrophe. This has been speculatively discussed in connection with Seyfert nuclei and quasars [68]. If there are sufficiently many stellar encounters to make the system isothermal similar considerations also apply to star clusters.

III.b.4. Evolutionary Instabilities in Stars; Formation of Red Giants and Supernovae

Red Giants:

They are created when nuclear burning of hydrogen in the core is exhausted. The core contracts and becomes hotter, whereas the halo expands and becomes colder. The implosion is halted by the ignition of He-burning. Such instabilities occur several times in the lifetime of a massive star.

Supernovae:

The incomplete gravitational collapse of sufficiently massive stars ($10 \leq \frac{M}{M_\odot} \leq 70$; $M_\odot$ = solar mass) seems to be most likely described by our model. Evolutionary calculations [69] of single non-rotating stars in this mass region show that those stars develop into an "onionskin" structure with iron-nickel cores of about (1.4-1.6) $M_\odot$ surrounded by Si, O, Ne, C, He, and H shells. The subsequent evolution of the core is an implosion with photodisintegration of iron into alpha particles and neutrons and with neutronization (electron capture) processes. This evolution is usually insensitive to the nature of any envelope, because the implosion velocity is more rapid than the speed of sound [70]. For more physical insight into formation and evolution of supernovae see the review articles [71]. We consider the core of the massive star to be our statistical system which has reached a critical radius and particle number. The spherical vessel corresponds to the termination of the core by the Si shell, which may act as a heat bath. The core collapse is halted when the Fermi pressure of the neutrons dominates. The model investigated here describes this implosion phase with good quantitative agreement [39], however neglects the relativistic nature of the electrons and the weak and strong interaction processes. Therefore the bouncing of the core is beyond the scope of the model.

EPILOGUE

Some of the prospects, open problems, recent achievements and announced results of Thomas-Fermi theory should be listed at the end.

One of the not yet satisfactorily solved problems is the derivation of the Thomas-Fermi equation in the gravitational case, which is far less smooth and elegant than for ordinary matter with fixed nuclei. Further investigation should be devoted to the exploration of the phenomenon of negative specific heat and to the phase transition. Here numerical computations are less interesting than analytical results.

The occurence of negative specific heat in a model of free particles on a gravitating sphere with variable radius has been studied in [72]. The authors also suggest that the origin of most phase transitions may be located in negative specific heat elements at a microscopic level and apply their interpretation to a simple model of a chemical reaction. Also at Cambridge the consequences of the gravothermal catastrophe for classical systems were deepened: Isothermal axisymmetric equilibria of gravitating rods are investigated in [73], and self-similar solutions for a thermal conducting gas sphere in gravothermal evolution are considered in [74].

From this non-equilibrium computations the question arises what kind of kinetic equation emerges rigorously from the Thomas-Fermi scaling. The Thomas-Fermi limit for the time evolution of a Fermi or Bose system has been shown [75] to lead to the Vlasov-equation for certain conditions on the potential, which exclude the $1/r$-case. The Vlasov-equation was derived in [76] as the thermodynamic limit of the time evolution for a classical mean field system, with some weaker conditions on the potential than in [75]. That the quantum mechanical time evolution leads the same kinetic equation is not surprising, because the Thomas-Fermi limit combines a mean field limit with a classical limit. Recently [77] the conditions on the potential for the quantum mechanical problem treated in [75] could be weakened to the conditions for the classical problem treated in [76], however the $1/r$-potential is still excluded.

Progress in the ground-state Thomas-Fermi theory has also been achieved since the fundamental paper [17]. For neutral atoms the Thomas-Fermi interaction energy is shown to behave proportional to ℓ^{-7} for large separation ℓ of the atoms [78]. For neutral molecules positivity of the pressure and compressibility in the framework of Thomas-Fermi theory was proved in [79]. The Thomas-Fermi theory for atoms can be improved by adding to the energy functional the von Weizsäcker-term. Existence and uniqueness of solutions of the Thomas-Fermi-von Weizsäcker equation was proved in [80] and the fact that they minimize the Thomas-Fermi-von Weizsäcker energy functional. The authors also proved the existence of binding of two (dissimilar) atoms, which cannot occur without the von Weizsäcker-correction because of Teller's no-binding theorem.

Attempts have been made to develop a temperature dependent Thomas-Fermi theory for statistics different from Fermi statistics. For classical systems and for Boltzmann statistics a derivation of the Thomas-Fermi equation for the canonical ensemble and solutions for two models showing a phase transition are presented in [81]. Replacing in case of Bose statistics the Fermi function by a Bose function in the self-consistency equation for atoms might lead to a negative density near the origin caused by the -1 in the denominator. A derivation of a Thomas-Fermi equation for bosons is an open problem even in presence of suitably regular interactions. Bosons in an external field have been studied, however, with the help of correlation inequalities [82]. This model can be exactly solved and exhibits Bose condensation even in one dimension.

REFERENCES

Prologue

[1] W. Thirring: Lehrbuch der Mathematischen Physik, Vol.1, Klassische dynamische Systeme, Wien, New York, Springer (1977), Vorwort

[2] W. Thirring: Lehrbuch der Mathematischen Physik, Vol.4, Quantenmechanik grosser Systeme, Wien, New York, Springer (1980), Chapter 4.2, p.226

Chapter I.aa.

[3] F.J. Dyson, and A. Lenard: J. Math. Phys. 8, 423 (1967)

[4] F.J. Dyson: J. Math. Phys. 8, 1538 (1967)

[5] A. Lenard, and F.J. Dyson: J. Math. Phys. 9, 698 (1968)

[6] E.H. Lieb: Phys. Lett. 70A, 71 (1979)

[7] W. Thirring: Stability of Matter. Lecture given at the Erice Summer School. Vienna-preprint: UWThPh-80-26 (June 1980)

[8] E.H. Lieb, and W. Thirring: Phys. Rev. Lett. 35, 687 (1975)
E.H. Lieb, and W. Thirring: Phys. Rev. Lett. 35, 1116 (1975)
E.H. Lieb, and W. Thirring: in Studies in Mathematical Physics: Essays in the Honor of Valentine Bargmann, eds. E.H. Lieb, B. Simon and A.S. Wightman, Princeton, Princeton University Press (1976)

[9] E.H. Lieb: Bull. Am. Math. Soc. 82, 751 (1976)

[10] W. Thirring: Commun. Math. Phys. 79, 1 (1981)

[11] E.H. Lieb: Rev. Mod. Phys. 48, 553 (1976)

[12] W. Thirring: Acta Phys. Austr. Suppl. XV, 337 (1976)

[13] G. Rosen: SIAM J. Appl. Math. 21, 30 (1971)

[14] S.L. Sobolev: Mat. Sb. 46, 471 (1938)
S.L. Sobolev: Applications of Functional Analysis in Mathematical Physics, Leningrad (1950), Am. Math. Soc. Transl. Monographs 7 (1963)

[15] W. Thirring: Vorlesungen über Mathematische Physik T8: Quantenmechanik grosser Systeme. Vienna Lecture Notes (1975)
W. Thirring: Lehrbuch der Mathematischen Physik, Vol.4, Quantentheorie grosser Systeme, Wien, New York, Springer (1980)

[16] E.H. Lieb, and B. Simon: Phys. Rev. Lett. 31, 681 (1973)

[17] E.H. Lieb, and B. Simon: Adv. Math. 23, 22 (1977)

[18] B. Baumgartner: Commun. Math. Phys. 47, 215 (1976)

[19] E. Teller: Rev. Mod. Phys. 34, 627 (1962)

[20] F.J. Dyson: in Brandeis University Summer Institute in Theoretical Physics 1966, Vol.1, eds. M. Chretien, E.P. Gross, and S. Deser, New York, Gordon and Breach (1968), p.179

[21] A. Lenard: in Statistical Mechanics and Mathematical Problems, ed. A. Lenard, Lecture Notes in Physics, Vol.20, Berlin, Heidelberg, New York, Springer (1973), p.114

[22] P. Federbush: J. Math. Phys. 16, 347 (1975)
P. Federbush: J. Math. Phys. 16, 706 (1975)

Chapter I.ab.

[23] S. Chandrasekhar: Mon. Not. Roy. Astr. Soc. 91, 456 (1931)

[24] S. Chandrasekhar: Introduction to the Study of Stellar Structure, New York, Dover Publ. (1958).
First ed.: Chicago, University of Chicago Press (1939)

[25] L.D. Landau: Z. Sowjetunion 1, 285 (1932)
L.D. Landau: in Collected Papers of L.D. Landau, ed. D. ter Haar, New York, Pergamon Press (1965), p.60
[26] J.-M. Lévy-Leblond: J. Math. Phys. 10, 806 (1969)
[27] M.E. Fisher, and D. Ruelle: J. Math. Phys. 7, 260 (1966)
[28] I.W. Herbst: Commun. Math. Phys. 53, 285 (1977)

Chapter I.ba.

[29] J.L. Lebowitz, and E.H. Lieb: Phys. Rev. Lett. 22, 631 (1969)
[30] E.H. Lieb, and J.L. Lebowitz: Adv. Math. 9, 316 (1972)
[31] E.H. Lieb, and J.L. Lebowitz: in Statistical Mechanics and Mathematical Problems, ed. A. Lenard, Lecture Notes in Physics, Vol.20, Berlin, Heidelberg, New York, Springer (1973), p.136
[32] E.H. Lieb, and H. Narnhofer: J. Stat. Phys. 12, 291 (1975)
E.H. Lieb, and H. Narnhofer: J. Stat. Phys. 14, 465 (1976)
[33] M.A. Krasnosel'skij, and Ya.B. Rutickij: Convex Functions and Orlicz Spaces, Groningen, P. Noordhoff (1961)

Chapter I.bb.

[34] P. Hertel, H. Narnhofer, and W. Thirring: Commun. Math. Phys. 28, 159 (1972)
[35] B. Simon: J. Math. Phys. 10, 1123 (1969)
[36] P. Hertel, and W. Thirring: Commun. Math. Phys. 24, 22 (1971)
[37] D.W. Robinson: The Thermodynamic Pressure in Quantum Statistical Mechanics, Lecture Notes in Physics, Vol.9, Berlin, Heidelberg, New York, Springer (1971)
[38] D. Ruelle: Statistical Mechanics. Rigorous Results, Reading, New York, Benjamin (1974)
[39] P. Hertel, and W. Thirring: in Quanten und Felder, ed. H.P. Dürr, Braunschweig, Vieweg (1971), p.309

Chapter I.bc.

[40] W. Thirring: Essays in Physics 4, 125 (1972)

Chapter I.c.

[41] J. Messer: Z. f. Physik B33, 313 (1979)
J. Messer: in preparation

Chapter II.a.

[42] B. Baumgartner: Commun. Math. Phys. 48, 207 (1976)
[43] A.W. Roberts, and D.E. Varberg: Convex Functions, New York, London, Academic Press (1973)
[44] R.B. Griffiths: J. Math. Phys. 5, 1215 (1964)

Chapter II.b.

[45] H. Narnhofer, and G.L. Sewell: Commun. Math. Phys. 71, 1 (1980)
[46] G.F. Dell'Antonio, S. Doplicher, and D. Ruelle: Commun. Math. Phys. 2, 223 (1966)
[47] H. Narnhofer: private communication
[48] E.B. Davies: Commun. Math. Phys. 30, 229 (1973)

Chapter III.a.

[49] A.S. Eddington: The Internal Constitution of the Stars, Cambridge, Cambridge University Press (1926)
[50] V.A. Antonov: Vest. leningr. gos. Univ. 7, 135 (1962)
[51] D. Lynden-Bell: in Brandeis University Summer Institute in Theoretical Physics 1968, Vol.1, eds. M. Chrétien, S. Deser, and J. Goldstein, New York, Gordon and Breach (1969), p.1
[52] D. Lynden-Bell, and R. Wood: Mon. Not. R. astr. Soc. 138, 495 (1968)
[53] W. Thirring: Z. f. Physik 235, 339 (1970)
[54] E.B. Aronson, and C.J. Hansen: Astrophys. J. 177, 145 (1972)
[55] P. Hertel: Acta Phys. Austr. Suppl. XVII, 209 (1977)
[56] A. Pflug: Commun. Math. Phys. 78, 83 (1980)
[57] H. Satz: private communication
[58] T. Çelik, and H. Satz: Z. f. Physik C1, 163 (1979)
[59] R. Hagedorn: Nuovo Cimento Suppl. 3, 147 (1965)
[60] R.D. Carlitz: Phys. Rev. D5, 3231 (1972)
[61] N. Cabibbo, and G. Parisi: Phys. Lett. 59B, 67 (1974)
[62] D.J. Gross, R.D. Pisarski, and L.G. Yaffe: Rev. Mod. Phys. 53, 43 (1981)
[63] H. Narnhofer, and W. Thirring: Asymptotic Exactness of Finite Temperature Thomas-Fermi Theory. Vienna-preprint: UWThPh-80-1 (1980).
Ann. Phys. (N.Y.), to be published
[64] J. Messer: On the Gravitational Phase Transition in the Thomas-Fermi Model. Preprint (1980). J. Math. Phys., to be published (scheduled for December 1981)
[65] J. Messer: Non-Monotonicity of the Mass Distribution and Existence of the Gravitational Phase Transition. Preprint (1981).
Phys. Lett. A, to be published (scheduled for Vol.83, issue 6, June 1981)

Chapter III.b.

[66] A. Pflug: private communication
[67] D. Lynden-Bell: Mon. Not. R. astr. Soc. 136, 101 (1967)
[68] L. Spitzer, and W.C. Saslaw: Astrophys. J. 143, 400 (1966)
L. Spitzer, and M.E. Stone: Astrophys. J. 147, 519 (1967)
[69] W.D. Arnett: in Explosive Nucleosynthesis, eds. D.N. Schramm, and W.D. Arnett, University of Texas Press (1973), p.236
T.A. Weaver, G.B. Zimmerman, and S.E. Woosley: Astrophys. J. 225, 1021 (1978)
[70] W.D. Arnett: Ann. New York Academy of Sc. 302, 10 (1977)
[71] T.A. Weaver, and S.E. Woosley: Ann. New York Academy of Sc. 336, 335 (1980)
J.R. Wilson: Ann. New York Academy of Sc. 336, 358 (1980)
W.D. Arnett: Ann. New York Academy of Sc. 336, 366 (1980)

Epilogue

[72] D. Lynden-Bell, and R.M. Lynden-Bell: Mon. Not. R. astr. Soc. 181, 405 (1977)
[73] J. Katz, and D. Lynden-Bell: Mon. Not. R. astr. Soc. 184, 709 (1978)
[74] D. Lynden-Bell, and P.P. Eggleton: Mon. Not. R. astr. Soc. 191, 483 (1980)
[75] H. Narnhofer, and G.L. Sewell: Commun. Math. Phys. 79, 9 (1981)

[76] W. Braun, and K. Hepp: Commun. Math. Phys. 56, 101 (1977)
[77] H. Spohn: On the Vlasov Hierarchy. Preprint (1980)
[78] H. Brezis, and E.H. Lieb: Commun. Math. Phys. 65, 231 (1979)
[79] R. Benguria, and E.H. Lieb: Commun. Math. Phys. 63, 193 (1978)
[80] R. Benguria, H. Brezis, and E.H. Lieb: Commun. Math. Phys. 79, 167 (1981)
[81] J. Messer, and H. Spohn: in preparation
[82] J. Messer, and A. Verbeure: in preparation

Lecture Notes in Physics

Vol. 114: Stellar Turbulence. Proceedings, 1979. Edited by D. F. Gray and J. L. Linsky. IX, 308 pages. 1980.

Vol. 115: Modern Trends in the Theory of Condensed Matter. Proceedings, 1979. Edited by A. Pekalski and J. A. Przystawa. IX, 597 pages. 1980.

Vol. 116: Mathematical Problems in Theoretical Physics. Proceedings, 1979. Edited by K. Osterwalder. VIII, 412 pages. 1980.

Vol. 117: Deep-Inelastic and Fusion Reactions with Heavy Ions. Proceedings, 1979. Edited by W. von Oertzen. XIII, 394 pages. 1980.

Vol. 118: Quantum Chromodynamics. Proceedings, 1979. Edited by J. L. Alonso and R. Tarrach. IX, 424 pages. 1980.

Vol. 119: Nuclear Spectroscopy. Proceedings, 1979. Edited by G. F. Bertsch and D. Kurath. VII, 250 pages. 1980.

Vol. 120: Nonlinear Evolution Equations and Dynamical Systems. Proceedings, 1979. Edited by M. Boiti, F. Pempinelli and G. Soliani. VI, 368 pages. 1980.

Vol. 121: F. W. Wiegel, Fluid Flow Through Porous Macromolecular Systems. V, 102 pages. 1980.

Vol. 122: New Developments in Semiconductor Physics. Proceedings, 1979. Edited by F. Beleznay et al. V, 276 pages. 1980.

Vol. 123: D. H. Mayer, The Ruelle-Araki Transfer Operator in Classical Statistical Mechanics. VIII, 154 pages. 1980.

Vol. 124: Gravitational Radiation, Collapsed Objects and Exact Solutions. Proceedings, 1979. Edited by C. Edwards. VI, 487 pages. 1980.

Vol. 125: Nonradial and Nonlinear Stellar Pulsation. Proceedings, 1980. Edited by H. A. Hill and W. A. Dziembowski. VIII, 497 pages. 1980.

Vol. 126: Complex Analysis, Microlocal Calculus and Relativistic Quantum Theory. Proceedings, 1979. Edited by D. Iagolnitzer. VIII, 502 pages. 1980.

Vol. 127: E. Sanchez-Palencia, Non-Homogeneous Media and Vibration Theory. IX, 398 pages. 1980.

Vol. 128: Neutron Spin Echo. Proceedings, 1979. Edited by F. Mezei. VI, 253 pages. 1980.

Vol. 129: Geometrical and Topological Methods in Gauge Theories. Proceedings, 1979. Edited by J. Harnad and S. Shnider. VIII, 155 pages. 1980.

Vol. 130: Mathematical Methods and Applications of Scattering Theory. Proceedings, 1979. Edited by J. A. DeSanto, A. W. Sáenz and W. W. Zachary. XIII, 331 pages. 1980.

Vol. 131: H. C. Fogedby, Theoretical Aspects of Mainly Low Dimensional Magnetic Systems. XI, 163 pages. 1980.

Vol. 132: Systems Far from Equilibrium. Proceedings, 1980. Edited by L. Garrido. XV, 403 pages. 1980.

Vol. 133: Narrow Gap Semiconductors Physics and Applications. Proceedings, 1979. Edited by W. Zawadzki. X, 572 pages. 1980.

Vol. 134: γγ Collisions. Proceedings, 1980. Edited by G. Cochard and P. Kessler. XIII, 400 pages. 1980.

Vol. 135: Group Theoretical Methods in Physics. Proceedings, 1980. Edited by K. B. Wolf. XXVI, 629 pages. 1980.

Vol. 136: The Role of Coherent Structures in Modelling Turbulence and Mixing. Proceedings 1980. Edited by J. Jimenez. XIII, 393 pages. 1981.

Vol. 137: From Collective States to Quarks in Nuclei. Edited by H. Arenhövel and A. M. Saruis. VII, 414 pages. 1981.

Vol. 138: The Many-Body Problem. Proceedings 1980. Edited by R. Guardiola and J. Ros. V, 374 pages. 1981.

Vol. 139: H. D. Doebner, Differential Geometric Methods in Mathematical Physics. Proceedings 1981. VII, 329 pages. 1981.

Vol. 140: P. Kramer, M. Saraceno, Geometry of the Time-Dependent Variational Principle in Quantum Mechanics. IV, 98 pages. 1981.

Vol. 141: Seventh International Conference on Numerical Methods in Fluid Dynamics. Proceedings. Edited by W. C. Reynolds and R. W. MacCormack. VIII, 485 pages. 1981.

Vol. 142: Recent Progress in Many-Body Theories. Proceedings. Edited by J. G. Zabolitzky, M. de Llano, M. Fortes and J. W. Clark. VIII, 479 pages. 1981.

Vol. 143: Present Status and Aims of Quantum Electrodynamics. Proceedings, 1980. Edited by G. Gräff, E. Klempt and G. Werth. VI, 302 pages. 1981.

Vol. 144: Topics in Nuclear Physics I. A Comprehensive Review of Recent Developments. Edited by T.T.S. Kuo and S.S.M. Wong. XX, 567 pages. 1981.

Vol. 145: Topics in Nuclear Physics II. A Comprehensive Review of Recent Developments. Proceedings 1980/81. Edited by T. T. S. Kuo and S. S. M. Wong. VIII, 571-1.082 pages. 1981.

Vol. 146: B. J. West, On the Simpler Aspects of Nonlinear Fluctuating. Deep Gravity Waves. VI, 341 pages. 1981.

Vol. 147: J. Messer, Temperature Dependent Thomas-Fermi Theory. IX, 131 pages. 1981.